# The Story Within

# The Story Within

## Personal Essays on Genetics and Identity

EDITED BY **Amy Boesky**

The Johns Hopkins University Press

Baltimore

© 2013 Amy Boesky
All rights reserved. Published 2013
Printed in the United States of America on acid-free paper
9  8  7  6  5  4  3  2  1

The Johns Hopkins University Press
2715 North Charles Street
Baltimore, Maryland 21218-4363
www.press.jhu.edu

An earlier version of chapter 3 was published by Kate Preskenis in *The Gene Guillotine* (Present Essence Publishing, 2012). Chapter 12 contains excerpted material from *If a Tree Falls* and is adapted by Jennifer Rosner with permission of The Feminist Press. Chapter 16 is adapted from *The Still Point of the Turning World* by Emily Rapp, © 2013 by Emily Rapp; used by permission of The Penguin Press, a division of Penguin Group (USA) Inc.

Library of Congress Cataloging-in-Publication Data

Boesky, Amy.
    [Essays. Selections]
    The story within : personal essays on genetics and identity / edited by Amy Boesky.
        pages cm
    Includes index.
    ISBN 978-1-4214-1096-8 (pbk. : alk. paper) — ISBN 1-4214-1096-6 (pbk. : alk. paper) —
ISBN 978-1-4214-1097-5 (electronic) — ISBN 1-4214-1097-4 (electronic)
    1. Medical genetics.  2. Genetic disorders.  I. Title.
RB155.S985 2013
616'.042—dc23      2013001931

A catalog record for this book is available from the British Library.

*Special discounts are available for bulk purchases of this book. For more information, please contact Special Sales at 410-516-6936 or specialsales@press.jhu.edu.*

The Johns Hopkins University Press uses environmentally friendly book materials, including recycled text paper that is composed of at least 30 percent post-consumer waste, whenever possible.

# Contents

## Acknowledgments

This book has been the work of many hands. My primary thanks go to the writers for their dedication to furthering understanding about the complex connections between genetics and identity. I have deep admiration not only for their candor and commitment to exploring these issues but also for the care they've taken in bringing them to light.

My agent, Richard Parks, supported this project from its earliest days. Richard's blend of sagacity and patience is rare in any field, and I am truly lucky to work with him. Jackie Wehmueller at the Johns Hopkins University Press took up the volume and has guided it through multiple drafts with utmost care; no detail has been too large or too small to warrant her compassionate engagement. Along with her, Sara Cleary has helped in numerous ways to bring seventeen writers (not to mention an editor) into the same volume, if not onto the same page. Thanks are also due to Linda Strange for her careful and respectful copyediting and to the efforts of the design and marketing teams at Hopkins.

A creative nonfiction grant from the Howard Foundation at Brown University and support from Boston College allowed me critical time away from teaching and other duties to develop this project in 2011–12. The Institute of Liberal Arts at Boston College generously funded a (forthcoming) symposium to bring contributors together to discuss these issues in person. Jim Keenan and Cathy Read (both at Boston College) collaborated with me to propose this symposium and to secure funding for it. Their insights have been extremely helpful. In addition, the Jesuit Institute at Boston College co-funded a 2010–11 seminar with the ILA on genetics, power, and identity (co-chaired with Jim Keenan) that initiated my interest in gathering personal essays on this subject.

As I worked to put this collection together, conversations with colleagues at Boston College have been particularly helpful, both inside my department and beyond. I consider myself especially fortunate to teach and research in a university that not only respects but actively fosters interdisciplinary work. Moreover, I am lucky in the community of writers at Boston College with whom I share students, drafts, and ideas. Beyond this, my colleagues in the emerging Medical Humanities, Health, and Culture minor at Boston College have expanded my appreciation for the complex ways in which our culture represents health and illness. I value their insights enormously.

Like many who teach, I owe thanks to students as much as colleagues. Members of my creative nonfiction courses have helped me question and reconsider connections between form and content in contemporary memoir. I also owe thanks to my writing students at the Dana Farber, who remind me how many issues comprise a life, beyond the often-blurry categories of "health" and "illness."

Deepest thanks go to my family. In some sense, the essays in this volume are all family stories, at least in part, many involving deeply loved relatives no longer living, parents or siblings or children whose genetic conditions have inaugurated research and reconsideration for those of us left behind. In my case, it's my mother whose legacy has shaped me in this way. But I imagine many forebears for each of us as I read these essays and reflect on where we've been. I am keenly aware of them as I think about where we stand now, under those imagined branches whose filtered light, at times, is difficult to distinguish from their shade.

# The Story Within

# Introduction

Amy Boesky

...............................................................................................

The body is never a single physical thing so
much as a series of attitudes towards it.

LENNARD J. DAVIS,
"The End of Identity Politics"

This collection grew out of the conviction that personal narrative
reveals a great deal about the connections between genetics and
identity. As contributors, we share challenges and complexities associated with genetic conditions, as well as the hope that ongoing research
may provide better clinical options in the future. Further, we share a commitment to disclosing our experiences, improving ways that genetic identity gets talked about and understood by the wider public.[1]

Rapid expansion in the field of genetics has deepened misunderstand-
ings about the extent to which genetic information determines us. As
these essays reveal, no aspect of genetic identity is simple. Risk of predicted disease may be perceived as unbearable by one individual while
remaining manageable for another. One person chooses to learn all he
can about his genetic make-up, while another defers that information or
actively chooses not to discover it. Whether, when, and how to intervene
in a predicted genetic condition are all highly individual decisions, as are
determinations about whether and how best to involve family members,
immediate and extended. Our responses to these issues remain as individual as we are.

Yet the complexities surrounding these questions are often misunder-
stood. As a writer and professor of literature, I believe narrative helps to

expose the ways that genetics impacts an individual's nuanced sense of self. To that end, I asked contributors to write personal essays about what it has been like to live with information about their DNA, to make decisions based on that information, and to assess its influence on family relationships, especially those involving children and genetic legacy.

I came to this project after completing a memoir about BRCA1, the mutation (sometimes called the "breast cancer gene") that runs in my family. As I learned more about my own history, I became increasingly interested in the experiences of people living with other kinds of mutations. How much has information about genetics mattered for them, and what decisions have they made in response to it? I found several memoirs on these subjects (some by writers represented here), but I could not find what I was looking for—a collection of first-person narratives by people living with or facing a range of genetic conditions. Writing about a particular mutation tends to get clustered with other texts about the illness or condition to which it leads. What would happen if we wrote across conditions, considering the larger questions we face? Is there a kind of culture of foreknowledge we share, and if so, how is it constituted?

This collection explores that question by bringing together writers who have (or strongly suspect they have) mutations for a wide range of conditions, offering a kind of "personal genetics reader" that sheds light on the distinctions among us, as well as our connections. Over the past eighteen months, I reached out to individuals whose lives have been altered in significant ways by information about their DNA. Some have already worked actively to raise awareness about these issues; others are making this information public only now. In the process of working and talking with contributors, I was repeatedly struck by the extent to which our connections and identifications crisscross back and forth in complicated ways. Often, the person whose experience feels closest may be facing a markedly different condition. The most subtle characteristics connect us; the most nuanced experiences render us distinct.

We all believe this information matters—often in profound ways. Yet none of us feel that human beings are defined or determined by their DNA. When we consider who we are as people, we reflect on multifaceted characteristics: work, age, family and personal relationships. Ancestry. Philosophical and/or spiritual beliefs. In this sense we are, to quote Whitman, "large, [and] . . . contain multitudes." We work in diverse fields. Several of us are full-time writers or artists. One contributor, a 41-year-old twin with cystic fibrosis, works as a genetic counselor. Our careers span the fields of consulting, academia, journalism, advocacy, and filmmaking. The youngest contributor, just out of college, is 22; others are in late middle age. Some are married, some single; some of us have children, some, given a known genetic mutation, have chosen, in the words of Anabel Stenzel, to "leave no child behind." Some contributors live with families; one lives in a Benedictine order. Though most of us have written about genetic subjectivity before, many of us found it difficult to go over this terrain again.

Despite this, contributors have written these essays with several common goals: first, to reveal the depth and complexity of lived experience associated with genetic conditions or illnesses; second, to de-stigmatize genetic difference, helping to demonstrate how widespread and varied these experiences are, even as they remain largely invisible; and third, to suggest that genetics poses a new frontier for us all, necessitating new decisions and determinations. Whatever lies ahead, our sense is that deeper and more sustained conversations about these issues will be more important in the coming years, and not less so.

## Finding Out: Genetics and Ideas of Self

Contributors focus on different aspects of genetic identity in their essays, which I have organized into three parts: first, *knowing* (has awareness or fear of a genetic mutation shaped your sense of who you are, and if so, how?); second, *intervening* (if you have a mutation, have you done something about it, and if so, what?); and third, *passing down* (has this information affected your feelings about family, children, and genetic legacy?). There are important connections among all three parts, of course, as these issues are intertwined, three strands of a single braid. For writers addressing the first question, identity is by no means stable, essential, or identical to genetic "status." There is a crucial gap between *information* about a genetic mutation (obtained from testing) and *knowledge* of it, and the latter is constructed differently for each of us. In this first part, contributors are still debating whether to learn about a likely mutation. Some believe that deferring results is a valuable option—not a decision against finding out, but a decision in favor of deliberate waiting. All five of us writing here believe there are crucial aspects to identity that transcend the results of a genetic test. Two contributors (Kate Preskenis and Charlie Pierce) have taken part in pioneering research on Alzheimer's—one as a subject, the other as an investigative journalist. Another (Alice Wexler) has a sister, Nancy, who led the team that identified the gene for Huntington's disease in the 1990s. For several contributors, questions about whether and when "to know" remain unresolved. These first essays (while focused on different conditions) interrogate the idea of genetic "knowledge" as they explore decisions surrounding what it means to "find out": hesitation, deferral, and recognition of the gaps between information (which a test can provide) and knowledge (which it may not).

"We never *see* this stuff!" doctors have said to Joanna Rudnick, the documentary filmmaker interviewed in the second part of this collection. They are referring to private conversations revealed by Joanna's documentary (*In the Family*) that take place before and after the brief moments a doctor spends in an exam room with women positive for a BRCA mutation. In such discussions, women debate whether and when to pursue surgery and what the mutation means for them in the future. They talk intimately about what it feels like to learn the results of genetic testing alongside close relatives—sisters, cousins—who may be obtaining results at the same time. Often, these conversations take place behind closed doors. What difference would it make if doctors, genetic counselors, or policy makers had access to what they say?

To my mind, there is a wide need to "see this stuff" and hear these stories. We are all affected by genetic information, which will become increasingly evident in the coming decades. If we ourselves are not personally touched by it, someone close to us will be. Yet most people have no access to these stories. Those of us with known or likely mutations may be sworn to secrecy by family members. Even if we are free to talk publicly, we may worry how others will perceive us if we disclose personal risk. Lingering stigma about genetic difference compounds the tendency to stay silent about a genetic history or future, to keep the details private.

Not hearing these stories, it becomes easy to judge or presume how much genetic information matters. Whether it is critical to have genetic testing (or why an individual might desist or defer). What informs the ethics of discovery and disclosure. What constitutes intervention, whether it is requisite, justified, or problematic. What guides the personal ethics of reproductive intervention. Not hearing these stories, some people decide genetic mutations are "over-hyped," that, fundamentally, they don't matter. At the other extreme, people may sensationalize genetic difference, feeling that people with mutations are different from everyone else. This happens even inside families. (One contributor said that the phrase assigned to family members who carry her family's mutation is "you people.")

Without access to individual experiences, we are more likely to accept the crude terms available to us ("mutant," "carrier") and less likely to challenge metaphors that estrange genetic identity, associating it with deviance, culpability, or malfunction. Wider dissemination of genetic narratives may alleviate the tendency to represent people with genetic mutations as somehow alien or "other" or, more subtly, to make these narratives conform to extant templates.

Not long after my book came out, a local bookstore owner asked me about responses from readers. I told her I was surprised how many emails I'd gotten from people facing hereditary issues unrelated to breast or ovarian cancer. These responses tended to be strongly empathic—"I know just how you feel," or, "I went through almost the exact same thing." Were there characteristics common to those of us facing hereditary conditions?

The bookstore owner seemed skeptical. "Honestly," she said in a conspiratorial tone, as if she and I were colluding. "Don't we all 'have' something? How different is it, really, having a genetic mutation, when in the end, we're all mortal?"

Her question stayed with me. In the largest sense, her point is right. We are all mortal. Eventually, most (if not all) of us will be found to have predispositions for one condition or another. But for now, this kind of foreknowledge is not evenly distributed. Her implication that I was overemphasizing the importance of known genetic mutations goes to the core of a debate currently being waged at many levels—among geneticists, historians and critics of science, and scholars working in disability studies and medical humanities. How much does genetics matter to the ways we understand ourselves, our relations to culture and family?

In my own case, the BRCA1 mutation has mattered profoundly. Hereditary cancer has shaped my identity as much (say) as being female or being a mother. Has it mattered? Yes. We all "have" something—and down the road, more and more people will discover that such "having" is tied to subtle variations in their DNA. But at present, genetic subjectivity constitutes a different kind of "having" than other kinds. Not necessarily greater or more serious, but different. Genetic subjectivity challenges stable, binary constructions of "health" and "illness." It comprises a kind of third position, an in-between state that may affect how an individual experiences time, how he or she thinks about the past and future. It may inform decisions about when (and whether) to have children or to marry. It may influence vocation, where and how to live, relationships.

In part I of this volume, contributors explore the unstable borders between identity and genetics from different angles. How is risk experienced? How deeply is a person's sense of self shaped by what is known (or not) about his or her DNA?

Most of us have understood for some time that significant conditions run in our families (conditions that vary in important ways, both in the trajectory of the associated disease and in available treatment options). What has changed since the discovery of genetic bases for these conditions (along with tests that forecast our disposition to develop them) is the specificity of the information we may now hold. What was previously understood as familial risk ("this runs in my family") is now experienced as individual certainty ("if I have this mutation, my chances of acquiring this disease may be as high as 80% to 100%"). Statistically, this information will affect some family members and not others. With certain conditions, such as Huntington's, the degree of risk can be closely predicted. In other cases, the predictive value of testing may be less certain.

Francis Collins has suggested that the desire to learn about genetic mutations stands in mathematical relation to the possibility of doing something about it. He calls this the RBI rule; "Desire to Know = Risk × Burden × Intervention."[2] Collins translates the third factor in his equation, "intervention," as a simple question: "What can I do about it?"[3]

For these writers, no equation is adequate to represent the dilemma of whether to "find out." Kelly Cupo—the youngest contributor here—sets her deliberations about testing for Huntington's disease in an essay describing suspended travels abroad during her junior year in college. The volcano in Iceland had erupted for the second time in weeks, halting air travel and leaving Kelly stranded. In an Italian bar, a group of guy friends began talking with her about how to choose an "ideal" mate. For Kelly, who had known for several years at this point that her mother has Huntington's, staying silent in this context was as difficult as speaking out. What does a 50-50 chance mean when you are trying to plan your life? Back in Connecticut, as Kelly describes it, a framed map on her family room wall charted her mother's own travels when she was young. The metaphor of the map opens a deep irony in Kelly's essay: where is the directive to follow when her own course remains uncertain?

Kate Preskenis, a member of a family deeply affected by early-onset Alzheimer's disease, is still ambivalent about "finding out" whether she carries the mutation that causes this condition. Now in her thirties, she describes her conflict about whether to learn the results of genetic testing she underwent several years ago as part of a National Institutes of Health study. For Kate, whose mother and aunt began to show symptoms of AD when they were not much older than she is now, fear of this mutation has shadowed many aspects of her life, including decisions about having children. Kate and her family members remain deeply invested in participating in research, believing that such participation may help to halt the course of this disease, but at the same time, Kate is cognizant of the "collateral damage" caused by a hereditary condition.

In "Driving North," journalist Charlie Pierce, who has written widely both on sports and on politics, describes a visit to Washington University during which he explored pioneering research on Alzheimer's, which robbed his father of his memory long before it killed him. AD is widespread in his family, and Charlie, nearing 60, reconsiders in his essay what memory means as we grow older and how the past weighs on what we face in the future. For Charlie, as for Kate, day-to-day missteps (forgetting a name, substituting a semicolon for an apostrophe) get experienced as "symptoms," suggesting a residual cost to living in the shadow of a genetic disease.

My essay, "In Samarra," situates my ambivalence about genetic testing for BRCA1 inside inherited family stories, looking at the forms of narratives that have shaped my family's understanding of risk, heredity, and cancer for generations. We pass down ways of representing and understanding information as much as we transmit our DNA. How do family myths get woven through other forms of knowledge, and how do we shape and reshape stories to make sense of what we know and what we don't? My family has adopted a narrative of medical intervention to obviate the fear of fate, yet our earlier "genetic" narratives continue to haunt our ways of knowing.

Last in this part of the book, Alice Wexler, whose sister helped identify the gene for Huntington's, looks back through the lens of history to explore what "knowing" and "not knowing" about HD has meant for her over the past four decades. Alice challenges reductive genetic determinism by offering insights into the range of issues we still don't understand about the epigenetic factors that contribute to the onset of HD. Alice's essay, "The Unnumbered," offers a history of a family that has changed the terms of knowledge about HD for the rest of the world. Yet, even for this family, the space of not knowing has important value.

## Intervening: Living with Genetic Difference

How does life change for people who know they carry a genetic mutation? What steps are taken to reduce risk, participate in research, or become advocates for others living with or facing the same condition? More broadly, how do ideas of genetic difference expand current ideas of culture and community?

"Doing something" about a mutation can be subtle as well as explicit. Many of us think about intervention primarily in terms of medical intervention, which may involve surgery, chemoprevention, participation in clinical trials, or closely monitored observation. Today, surgical procedures are possible that did not exist a generation ago. The Stenzel twins, Anabel and Isabel, who co-author an essay about living with cystic fibrosis, describe the changes they have seen in terms of medical options over the past several decades. At 41, they feel that it would have been impossible to imagine, years earlier, what they have been able to accomplish. Mara Faulkner, who has hereditary retinitis pigmentosa, a condition leading to late-onset blindness, also notes the value of new inventions such as canes with embedded GPS technology to help those without sight. For women living with BRCA mutations, as my interview-essay with Joanna Rudnick makes clear, preventive surgeries or chemoprevention may allow decades of life without disease.[4] Medical interventions (and technological advances) have been game-changers for a number of us. In the final part of the volume, Jennifer Rosner writes about the opportunities such interventions have offered her daughters, born genetically deaf, and Laurie Strongin describes a decade-long battle to save her son, born with Fanconi anemia, through pioneering reproductive technologies.

Yet, many of the conditions written about in this volume cannot (yet) be prevented or alleviated by medical intervention. In some cases, clinical care may only marginally prolong life or reduce suffering. Not only is treating a disease markedly different from intervening in the hope of preventing it, but in some cases, as Emily Rapp notes, the issue is not about "repair" at all. Not all writers here equate intervention with clinical care or surgery, nor do all see medical intervention positively. The idea of "medical repair" may be problematic for individuals who challenge our culture's understanding of disability and bodily difference. On a more mundane level, medical intervention (if available) does not always turn out as it should. Preventive surgery can extend or even save lives, but it can also be ill-advised, the "repair" worse than the condition it intends to heal.

Intervention—what Collins says translates to a simple question ("What can I do about it?")—is not always simple. Moreover, even when there is nothing medical to be done, there are important nonmedical ways to do something about a mutation—from education, outreach, and advocacy to writing, filmmaking, and fundraising, enlarging what Mara Faulkner refers to as the "community of compassion" to make the lives of *all* people better, those living with known genetic mutations and those who are not (or not yet). Many of the conditions written about in this volume are associated with foundations and organizations that offer much-needed resources and information; in fact, several contributors have either begun or advocated for these foundations. All of these organizations could benefit from greater exposure and support, and writing or fundraising on their behalf constitutes one form of intervention.

Public disclosure is another. Most of the writers in part II emphasize the need for stronger and more effective communication about genetics, such as Misha Angrist's call for better mechanisms for obtaining informed consent. In his work at Duke, both as a professor and in reviewing genetic research protocols, Misha details the need for greater transparency about genetics, reminding us that more is to be gained than lost from such disclosure. Misha anticipates a day when genetic information will be seen as just one more piece of information—like cholesterol screening or blood type. In pressing for such demystification, Misha describes his decision to take part as one of the original "ten" subjects in publishing his genome with George Church's Personal Genome Project. Misha sees the mystification of genetic information as dangerous in both the short and the long term; disclosure, he suggests, is a critical first step in changing popular misconceptions of what mutations "mean."

After Misha, several writers consider the potential hazards (Michael Downing) and benefits (the Stenzels) of medical intervention. In "Help Wanted," Michael writes about learning he has a mutation for hereditary coronary disease, which causes enlargement of the heart (a condition whose first symptom, he wittily observes, is death). The negative consequences of surgical intervention have carried great costs for Michael and his family, physically and emotionally. Paradoxically, for Michael, it was not surgery that provided real intervention but the more subtle act of reclaiming his own narrative from a medical master plot dictating what was wrong with him and how to "repair" it.

A different view of medical intervention is described by twin sisters Anabel Stenzel and Isabel Stenzel Byrnes, whose experiences with cystic fibrosis and organ transplantation have led to writing, filmmaking, and international advocacy. The Stenzels write about being twins, how CF has impacted them, and how medical intervention has extended their lives in ways formerly deemed impossible. Their co-authored essay makes clear the extent to which intervention re-informs their combined sense of writing and outreach. Not only do Isabel and Anabel see themselves as profoundly "twinned," but they recognize that living with CF has always been, for them, about pushing back boundaries, working for *more*—more time, more health, more education and improved quality of life for others living with CF.

How can talking with others about a genetic mutation constitute a kind of intervention? Patrick Tracey, whose family has been affected for generations by hereditary schizophrenia, sees disclosure and dialogue as critical vehicles for demystifying once-secret familial mental illness. Patrick's essay looks back, to his research on the roots of his family's schizophrenia in Ireland, as well as forward, to wider-ranging conversations with families in America. Patrick found to his frustration that hereditary schizophrenia in Ireland remains a topic that cannot be raised (or explored). Yet he believes what cannot be talked about *must* be talked about, as intervention demands beginning or extending current conversations about genetics and mental illness.

The idea of disclosure takes on different resonance for a filmmaker. Joanna Rudnick is committed to making genetic stories visible, both in her documentary about the BRCA mutation and in her new project on photographer Rick Guidotti and genetic variation.[5] Rudnick describes the need to bring family stories into the open, primarily by "inviting the audience to see" moments of difficult decision making that shed light on social and cultural issues surrounding genetic mutations. Looking in—and looking on—is essential to "steady the gaze" (a phrase Joanna borrows from Rick) in order to accept genetic difference without looking away.

One hope is that such disclosure will widen our sense of what community means and how we see each other. Mara Faulkner's essay "Community and Other Ordinary Miracles" challenges deeply held ideas of self on which current discussions of genetic identity depend. Mara, who has hereditary retinitis pigmentosa, has lived in a Benedictine order for most of her adult life. She sees the "ordinary miracle" of communal life— and a re-envisioned community of compassion—helping to mitigate the potential challenges of blindness. Mara's essay examines the ways that blindness is represented in our culture, how deeply entrenched our metaphors remain for sight and understanding, and how limited our capacity is to participate in the "ordinary miracles" that could extend compassion to those who live (and experience) the world beyond narrow and restrictive norms.

## Passing Down: Genetics and Family

How do known or suspected genetic mutations affect our understanding of family and the future? In this last part of the anthology, the awareness of nonlinear aspects of transmission is especially pronounced. Children change their parents, as much as the other way around. The child is father of the man, as Wordsworth wrote, and while mutations can be passed down, knowledge of them can be and often is "passed up."

Most of these contributors learned they (or their partners, or both) had a genetic mutation only after the birth of an affected child. For Jennifer Rosner ("String Theory"), previous scholarship in philosophy on the "messy" borders of self-identity was revived when her first daughter was born deaf. Advocating first for this child, and later for her second daughter—born profoundly deaf—led Jennifer to confront difficult (and urgent) choices. Should she and her husband augment their daughters' capacity to hear—in one case, with technology, and in the second case, through surgery? How do you make choices for a prelingual child? Jennifer's essay explores the divisions she confronted in our culture while navigating these choices—the eagerness of strangers to offer advice; the polarization over the use of hearing aids or cochlear implants in Deaf children; the hermeneutics of "listening," often lost in the noisy clamor over what hearing means and how it should be attained. Jennifer's daughters, now school-age, have helped to extend her sense of what listening means and how it may be experienced in ways that expand our previous conceptions.

In "The Long Arm," Clare Dunsford describes a previously unknown aspect of the mutation for fragile X that she passed down (before knowing about it) to her son, J.P. Clare's essay reminds readers that what we think we know about a genetic condition is often still in the process of unfolding. Fragile X, as it happens, carries a "rebound effect," as researchers are now learning; carriers—long thought to be unaffected by the premutation—are now being found to develop neurological symptoms late in life. How we experience time, as Clare's essay makes clear, is altered by such a premutation, for time can move in circles, backward as well as forward, up as well as down. What we thought was over may in fact be only the beginning of something more challenging. Here, the braided aspect of identity is revealed in all its complexity. We are back where we began: What does it mean to know? And what will you do about it?

The tie between intervention and heredity is explored in "Lettuce and Shoes," where Christine O'Hagan writes about the experience of

Duchenne muscular dystrophy in her family. Having lost two uncles and a younger brother to DMD, Christine was still unprepared when Jamie, the second of her two sons, also turned out to have DMD—as did both of her sister's sons. While ordinary life goes on, and the writing life has provided a much-needed emotional outlet, Christine has been deeply altered by this familial disease. Years later, she still relives the pain of watching Jamie suffer, while she continues to feel profound appreciation for his life.

Laurie Strongin, whose son, Henry, died ten years ago of a rare genetic disease called Fanconi anemia, has become an advocate for pre-implantation genetic diagnosis, which—in the right circumstances—can enable families with affected children to intervene in their children's genetic health prior to embryonic implantation. The Strongins struggled for many years to use PGD to have a child who, as a perfect HLA match, could probably have saved Henry's life through bone marrow transplantation. Laurie recounts the story of her family's struggle to keep Henry alive and her continued advocacy on behalf of PGD ten years after his death. The foundation set up to honor Henry's legacy has helped other children and will continue to do so. Laurie's essay on PGD offers a new view of genetic legacy—a window into a future where it may be possible (and perhaps one day more affordable) to select against genetic disease prior to birth.

In the last essay in the volume, "Dear Dr. Frankenstein: Creation Up Close," Emily Rapp (who herself has a disability) writes about giving birth to a baby boy with Tay-Sachs, a neurological disease that is always terminal. Having lost a leg in early childhood and recognizing some of the challenges associated with living with a disability, Emily was determined to seek prenatal testing when she learned she was pregnant. But the preliminary screening she underwent for Tay-Sachs missed Ronan's rare mutation. Ronan's life altered every plan and idea Emily once held. She explores her experiences with Ronan through her complex identification with the novel *Frankenstein*, sympathizing, at various points, with Mary Shelley, whose unlikely ghost story troubles our understanding of what it means to risk "creation"; with Dr. Frankenstein, ambitious and at times thoughtless creator; and finally, with the creature himself, cast out by a world that finds him unbearable.

Writing collectively about these experiences has reminded many contributors that the current vocabulary for talking about genetic difference is inadequate at best. Words like *mutant* feel offensive, *carrier* suggests contagion, and many other terms associated with genetic identity seem pejorative.[6] Lacking appropriate terms for discussing genetic identity, we sometimes fall back on disease models—people talk about "being diagnosed" with a mutation, for instance, or "having" a gene such as BRCA (rather than inheriting a mutation in that gene). What has been called the "abjection of illness" here translates to the abjection of genetic difference.

This abjection is shored up by the subtle (and insidious) use of metaphor. Susan Sontag argues that metaphor works to estrange and mystify the experience of illness, rendering it as less real and thus all the more frightening.[7] To my mind, our current genetic metaphors work to mystify and estrange difference along the lines Sontag suggested for TB and cancer in earlier periods. I find two kinds of metaphor for genetic mutations especially prevalent among science writers and journalists. Metaphors of deviance or culpability suggest the presence of an internal criminal that must be identified and policed: genes are referred to as "bad," "rogue," "lurking," or "culprits" that need to be hunted down, exposed, and classified.[8] Metaphors of malfunction describe mutations as "errors," "flawed copies," damaged, or broken, enforcing a mechanistic approach to heredity that Mary Rosner and T. R. Johnson see as shoring up rigid ideas of the body as "machine."[9] Both metaphors—the deviant and the mechanistic—focus on the mutation in isolation, with no reference to the human being of which it is a part (and only a part). As these metaphors perpetuate residual stigma associated with genetic difference, they dehumanize us all. I believe, to follow Sontag's terms, that such metaphors need to be "exposed, criticized, belabored, used up,"[10] and I hope the candid language of these essays will be an important step in that project.

Memoir—the narrative exploration of self—is often organized around a central change or transformation. Yet genetic memoir—and I use this term deliberately, as I think these essays represent a new turn in the narrative representation of self—treats *change* in distinctive ways.[11] Stories about genetic susceptibility do not conform to existing models, either for writing about illness or for writing about selfhood. Arthur Frank, writing on disability studies and illness, has objected to the reductive power of what he calls the "restitution" narrative, which he sees as our culture's preferred way of writing and reading about disease. In such narratives, the protagonist gets sick, gets treated, and gets well again, often ending up "better" than before.[12]

The individual living with a genetic mutation has no access to the banal heroics of restitution narratives, for the mutation story is not that story. People with inherited mutations don't and can't "get better." Instead, many of us with suspected or known mutations live in anticipation (or dread) of what may be—not so much what Frank calls the "culture of remission," but instead a culture of "pre-mission." Such "pre-mission" shares some of the liminal qualities Laura Tanner perceives in medical waiting rooms, spaces in which the "cultural categories [Lennard J.] Davis describes as 'immutable' are fraught, contested, or blurred."[13]

For writers of genetic memoir, such liminality is evinced in several ways. First, disclosure is problematic. It is difficult to reveal truths about your genetic status in the light of the anxieties and prohibitions of other family members, some of whom may have strong convictions about the need for privacy. All memoirists worry about disclosing family secrets (about alcoholism, for instance, or abuse, or mental illness), but genetic memoirists have a particular concern, because what gets disclosed about their own DNA affects family members directly. "I am acutely aware of what I can and cannot say about my siblings," one writer admits. Some writers share drafts of what they've written with family members; others agree to make certain topics off limits, or restrict themselves to preapproved terrain. But even when permission is granted, the genetic memoirist is aware she is writing a story with blurred boundaries; she may feel she is writing her sisters' story, or her mother's, or that of her children, as well as her own. Writers use different strategies to expand traditional, "bounded" ideas of authorship. The Stenzel twins co-author their writing on genetic identity, often choosing the pronoun "we" over the more conventional first person (while still retaining strong individual voices). Other writers include dialogue in their essays or interview family members (or doctors) to expand the single "voice" of conventional first-person authorship. Genetic memoirs explore new ways to expose identity as layered, lyrical, and highly relational, a hybrid of self and other, "I" and "we."

In addition, the genetic memoirist challenges conventional plotlines. He or she is acutely aware of temporality in writing and may work to find strategies to blur the linear progression of much first-person narrative. He may sense that he is writing backward as well as forward. For many, a genetic condition becomes associated with a certain age—perhaps the age at which a parent first showed symptoms or became ill. Sometimes, as Alice Wexler points out, this manifests itself in a preoccupation with numbers: CAG counts or repeats; blood markers in a CA125 test. Many of the writers here talk explicitly and repeatedly about age: this is how I felt at 22; at 37; at 53. Some writers, frustrated by linear plotlines that presume a one-directional motion, disrupt temporal movement through flashbacks, dreams, or moments suspended in time (like the scene in the bar in Kelly Cupo's "Undiagnosed"; the dream of the autopsy in Kate Preskenis's "Collateral Damage"; the twilit dread of reduced sight for Mara Faulkner; the wintry half-dreams Emily Rapp experienced while pregnant at Yaddo).

Resistance to linear narrative is often allied with grief; for the parent in mourning, the "now" of the present is shadowed by earlier loss or looming tragedy. Laurie Strongin's "now" is the ten-year anniversary of the loss of her son; for Christine O'Hagan, the memory of the son and nephew she has lost is present in every minute of every hour. These are not griefs to be overcome or put away, nor can they be represented as such. Future loss casts a shadow over the present. Even as she moves away from the acute grief of losing her son, Christine grieves in advance for her dying nephew. For Emily Rapp, the "now" of her son Ronan's life is lit up by awareness of his impending death with brutal, solar clarity.

For many of these writers, the past gets revisited as a traumatic site, mined and re-mined as they struggle to make sense of it. This is another distinguishing feature of genetic memoir. Many of these writers structure their essays around repeated tropes, images, or stories. Such repetition, as trauma theorists note, is part of the feeling of being caught inside an experience that re-injures.[14] Such repetition may be represented by a literal return—Patrick Tracey, for instance, goes back to ancestral sites in Ireland, trying to understand the source of his family's heredity mental illness. Or it may be a figurative return, traced and retraced through memory. To remember the disability or death of forebears from a condition you face yourself is to cycle between present risk and past grief. In this way, the genetic memoirist moves between past and present, woundedness and recovery.

Some critics have objected to what they see as "conflicting paradigms" in memoirs about disability and bodily difference. C. Thomas Couser, for instance—even as he admits such narrative has "considerable potential [for] counter stigmatizing"—suggests that "disability memoir" may fail to challenge the ways our culture shores up existing ideas of what is considered healthy, beautiful, or normal.[15] I believe the writers here are working, in different ways, toward the emancipatory rhetoric Couser urges. Rather than "triumphing" over a genetic mutation, or "repairing" it, or being rehabilitated from its effects, these writers challenge readers to reconsider the "physical, social and cultural obstacles" associated with genetic difference.[16]

The field of genetics is changing rapidly. We expect, in years to come, there will be more information available to us all and, hopefully, new treatment options. Beyond expanding information, however, we hope to see increased awareness of the terms in which such information gets conveyed. We hope for a wider and more capacious understanding of human variation, not only in terms of biological makeup, but in terms of how that variation gets represented and understood.

## PART ONE

# Finding Out

*Genetics and Ideas of Self*

# ONE

# Undiagnosed

Kelly Cupo

..................................................................................................

*In April 2010, the Icelandic volcano Eyjafjallajökull erupted for the
second time in less than a month, pouring floodwaters from its glacier
and releasing a cloud of ash that moved across Europe. There are
probably lots of volcano stories out there. Here's mine.*

Once upon a time, sophomore year of college, I was sitting on my bed
talking to a friend—we'll call her The Friend—about a former roommate—we'll call her Rachel—who wasn't delivering on the friendship
part of our friendship. Run of the undergrad mill. The Friend was sitting
on my roommate's roll-y chair. Me, under my covers in bed.

We'd just gotten back from breakfast in the dining hall. Pancakes
like paper. Butter, unwhipped.

"I feel like Rachel's being really selfish, and it's ticking me off. She
comes in here and is friends with me when it's convenient, but then she
turns it off in a second when it isn't," says yours truly: heroine, princess. Me.

"Hmmm. Well," The Friend says.

"I get that we're not totally compatible, but I really care about her!
We lived together a whole year, and I just don't think I should be treated
that way."

Her turn. "You know what, Kelly, I feel like you get extra-sensitive
when it comes to thinking people are selfish. It's probably because of
your mom and everything."

Oh what.

We were not talking about my mom. Why are we talking about my
mom? She goes on. She's diagnosing me.

"Really, Kel, we're all kind of selfish." Wait, are we talking about me now? Oh. She's not waiting.

Spins on the roll-y chair. Stares at me with dark dark eyes. "You're selfish too, you know. You're here at college when your mom is at home sick. Instead of being there for her, you're here, hours away. In *Boston*." Pins me. To a place I'm not ready to be.

Her eyes were like black holes. In I went. This was a close friend. This was a tough subject. I didn't have the perspective. From inside the black hole. That I should feel any different.

Black-hole Kelly knew studying abroad was selfish. If Boston was far away, what was Europe?

But I couldn't not go.

Once upon a time, my mom traveled.

On the walls of our TV room back home, we have a map hanging on the wall of where she went. Made by my grandma Marcia, framed in plastic. Matted off-center. Casual. It hangs right there next to the TV we used to watch like a god.

I grew up with that map. Cartoon clippings about aching feet, a dotted Wite-Out line outlining my mother's stops. She cut out little pictures of clocks to show where she went when. A miniature Big Ben, the clock hands hard to read. The map is about the same size as the TV next to it. They L each other. That's how I planned my trip: from my mother's map. From TV.

In my plans, there was no volcano. No ash and no in-between.

Abroad, I felt closest to my family. I even thought I liked the parents at some points. (A feeling remedied by my three-week stint back at the casa de Cupo upon my return. And by parents I mean my dad. Because my mom fell off that peg a while ago. Since she's sick. But she's "proud, so proud" of me even so. I give her that.) Abroad, let's suppose the parents were still a duo. Let's just say we connected. Maybe because they were 20,000 leagues across the sea. Or maybe because they were completely behind my cause for the first time. They wanted me to have lush and luxe adventures. Do everything. My dad gave me advice: *hitchhike if you miss the train.* I talked to them eight times as much. They told me things, and I listened a little.

For this, they had a map.

My mom always said her favorite place was Salzburg. We sang the soundtrack from *The Sound of Music* half my life. It's ridiculous to me that other people don't feel that music as a background overture to life in general, that other girls didn't act out the Liesl gazebo scene like I did, with my bedroom door closed and my nightgown on.

That was my mom. The way I miss her. Undiagnosed.

I am 22, and I'm a 100 percent undiagnosed Huntington's disease patient. The 100 percent glows like a film on my bod. I don't understand auras, but it disturbs me when people aren't aware of my 100-percent-ness. It angers me when they forget I'm a patient. When the person I'm in a relationship with knows I have "patient syndrome," they need to remember it. You can't just forget when you talk to me about things like "what I want to be." And I think, *Are you kidding me?* Isn't it obvious? I want to be a 100-percent-not-diagnosed-with-Huntington's-disease person.

*Look at that I'd be a person!* God said man should be free. I'd like to be that part of the God story. Without the constant break-up loop playing inside my head.

The loop where the person breaks up with the patient.

I know it will happen.

I tell myself, *It's not just about being a patient. Relax.*

I'm not worried someone won't take me to the doctor. It isn't that. They will take me.

Based on the look my doc gave me at my last physical, I have maybe ten years to live. So it's forward and back. Sometimes it takes me half a day just to get out of bed and realize: *You're currently pathetic. And this might be the end.*

I have been 0 percent diagnosed with Huntington's disease. My mom has been 100 percent diagnosed. And, god, is that not going well. If you think that's a complaint, read slower. Then slower and slower. Until you hear what I am saying.

Some of my favorite memories of my mother are of days she would skip work and I would skip school and she'd take me to yet another doctor to solve my unsolvable fifth grade stomachaches. She wouldn't take "just stress" for an answer, so neither would I. She'd bring me to McDonald's for breakfast after. And I'd get pancakes (they called them "hotcakes"). And she'd get an Egg McMuffin and a coffee. But sometimes she wouldn't get anything because it was late in the day and she'd already eaten, but if I had to get blood drawn, she said, I deserved a treat. I've never stopped liking whipped butter. Since I first watched it at those breakfasts, sliding right down the hotcakes with a calm, honest melt. At McDonald's we were free. It was in-between time, when you were 100 percent where you're supposed to be. You were accounted for. School knew. Work knew. But you were also 100 percent lost where no one could find you. Lost on purpose. Untracked. Together. That's what being a patient of Huntington's disease is like.

HD patients rip people they know apart. That's an early stage symptom. And not just people you care about. People who bring you to hotcakes. You know, *people*.

It goes on and on, this thing in my family. I get to decide. Do I want to know? If it were something else (say, depression), maybe I'd want to find out and they could tell me and I'd still be hopeful. I'd think (maybe), "They'll help me take care of it." Huntington's is different. It's untreatable. Un-treat-able. *No treats.*

What kind of disease is this? What they say when they tell you is they're "going to do their best to keep you comfortable" for ten to fifteen years. What are you supposed to do with that? What's really going to make it easier?

I happen to know what helps is the people around you. Except I don't really know that. Because right now I'm too scared to think about what I would say to them, the people for me. I know they would bring me to the doctor when I needed to go. Or out to lunch if I hinted. Which I would. But what am I supposed to say? I'm going to say, "It's been a pleasure being your friend, for this, this, and this reason." But I should say it right away because at some point (maybe) I won't be able to.

At some point I'll start talking about useless information that you don't really care about. Like my ability to eat chicken. And then my inability.

What kind of friend will I be then? I won't be able to enrich your life with new ideas, or support you deeply.

My mom says, "I'm so proud of you" about a job I don't even like. I miss her.

She used to know how to dream and how to tell me about her favorite movies.

About Salzburg and Liesl and the gazebo and "Climb Every Mountain" and the green-green hills.

People for me: they'll miss me the way I miss her. I'll be there, but I'll be lost.

It's a stretch. Can you picture me writing goodbye letters or thank-you letters to people when I get diagnosed? How obscenely self-centered that is. I'm alive. And will be for a while. But then. My whole body will untether itself from its normal responsibilities and functions. I'll become a diagnosed patient of Huntington's disease. Someone to track if you really want to. If you're into science, etc. Some will count their tracking as community service.

Can you picture me writing these letters.

Oh wait, you actually can. You actually think it's a *good idea*. They can read one at a benefit. For a charity, full of clapping people I don't know. Whatever, you will fucking miss me.

But missing someone, that feeling goes away. All those parts of your life where you made room for me will be vacant. But you'll fill them with something else. While I'm still alive. Because I won't exist anymore. Not that way. Not as the friend, the girlfriend, the sister. Not the wife. Not the mother.

Can you tell I don't know what I want. That I'm frozen. That I don't know what I should do.

I'm 22, and I'm unsingle.

I'm supposed to love me. Check. Love him. Check. Then no more single. It's math math math.

I wasn't always frozen. I used to have more search. I wasn't always "unsingle." I wasn't always a 100 percent undiagnosed Huntington's disease patient.

I wasn't always stuck on the word *chance*.

Like, *what are the chances?*

What were the chances that volcano in Iceland, named Eyjafjallajökull, would erupt for a second time the week I was trying to get back to home-base England during my junior spring abroad. From Rome. Ash clouds seeping through the atmosphere across Europe. Planes grounded. People stuck in between.

My junior spring.

I knew my mom was sick but I went anyway. I was 20. My mom had been sick for a long time, but I'd only known about the Huntington's for two-plus years (I hate counting).

Going abroad meant seeing things I'd always heard my mom talk about. Side trips to places like Salzburg.

There was a lot at stake. A lot of crazy I dragged overseas. A lot of black-hole feelings.

A friend of mine talks about something he calls the "redneck gene." He says he can't get serious with a girl whose family invests in cheap cars and vintage neon beer signs. It's a dealbreaker for him, he tells me. *Dealbreaker*. There's a word you hear too often, I think.

Just because you have a boyfriend doesn't mean you're "in a relationship." And just because you have a gene for Huntington's disease doesn't mean you're dead. It just means there's a good chance you could be soon.

What's a *good chance*?

Sometimes things look good if you step out of your life, look at it from the outside in.

Try picturing the way your latest relationship looks on a YouTube clip, or Facebook Timeline. Does it make you happy.

At the end, I'm stuck. Frozen in my gene pool. There's no "life event" for that. There's no "guess the fuck what, I'm gasping for air and I think you should all know about it" button, because what would we ever do with that button. Who would press it back?

Spring break, in Austria, I made my guy friends take a Sound of Music bus tour past vistas so green they made you feel you were gonna fall out the window and fly. My mom's favorite: "Climb Every Mountain." She could sing like nobody. Remember the time she was in *Anything Goes*? No wonder I cried on that bus. Those staggering green hills.

The three guys I was with didn't know about my soundtrack. They were there because I'd begged them. They were good guys. They had to give up a tour of Hitler's Eagle Nest with its armory castle to join me. The tour people played a video, some "Where Are They Now" clips between those shimmering Alpine sites. Liesl all grown up, starring in this or that. Staring out the window at those hills. I don't think the guys knew how hard I tried to keep from crying. One of them fell asleep during the ride. Sitting next to me.

When the volcano erupted for the second time, it was during the glam finale of our trip. It was April. I was sitting in Rome. Kicking back. Heating up. Scoping. Soak soak soaking it. But also sweating it. It was the end of spring break. After Salzburg and the bus tour I'd continued backpacking for two weeks and I was dirty. No matter how much I showered, the dirt sucked itself down into every crevice of the formerly beautiful Five Star pocket folder that kept my itinerary safe. I had my ticket all set for a plane back to London. Back to my life, my schedule. You know, finals.

Then, April 14, the volcano erupted.

Most volcano stories are, like, Oh my god! The fucking heavens are on FIRE. End. Rebirth. But this one wasn't. It was more like there I was in Rome, trying to eat food in ragu form. This volcano wouldn't exactly find you and burn you up. But it could give off ash that gets sucked into your airplane engine and eventually you'd go down for sure.

I had to get back to my exams, and papers, and then home—to America. The volcano stopped all that. Kept us there, under its smoggy load.

The reports came in. A giant plume of volcanic ash was moving south from Iceland, covering the British islands. Coming for ya. *Eyjafjallajökull.*

Eight hundred people had to evacuate the area around the volcano. Twice in a month, they said. *What are the chances?*

The American telecasters avoided saying the name of the volcano because they couldn't pronounce it. EYE-a-fyat-la-jo-kutl. Made it a deep trick, like Voldemort.

The American parents were scared for us. Mine texted me. Like a $3 text. The information was unclear: There was no knowing when the cloud would clear. Ash cloud. Swarms, seeping. Tiny granules of toxic rock. Were our parents really worried? just for us? What about the people in Iceland, didn't they worry about them? Were they serious or just 50 percent serious?

I was in Roma Termini, surrounded by desperate people, all trying to find trains because the planes couldn't fly.

I was thinking, this is the most anticlimactic volcano situation ever. I'm trying to find a train because I have finals to get back for? You must think I'm an asshole. Like I have all this time and money to spend getting back to London by land. For *finals*. Lots of people had bigger problems. The people near the volcano, for instance. People running out of money all over Europe. There was a recession. Everyone was on a downward escalator, and now there was this toxic cloud.

Stuck under this ash cloud, at first I was just dirty. Let's wait it out. Relax. Day 2, I reconsidered. We had no idea how long we could be stuck. Then my mind filled with heat. I tipped over. Out of my control. I couldn't afford to stay in Rome much longer. I couldn't afford train tickets to London. What was I doing studying abroad when I didn't have the money? What was I doing this far away? It escalated. "This will ruin me," I sobbed into my purple Razr phone to my father. Standing in a bakery of delicate Roman sweets. Why was I so absurdly unsettled? I was already ruined. And it wasn't the money.

I kept remembering a few nights back in a hostel bar in Salzburg. The night before the Sound of Music tour. I was with the boys. Me, Jason, Zack, Billy. In an open bar with tables and chairs that were too high and too big, like we'd all been shrunk down Munchkin size. *Climb every barstool.* We were drinking beer. Talking about relationships. Funny at first, then serious.

Billy started. Blond-haired Billy, electric-blue eyes, tall, thin, scruffy beard, straight jeans, Ray-Bans. "People say they don't want to settle down right now, but I actually do. I really just want a girlfriend," he said.

I left out his perfect jawline. (Is there a gene for that?) But he's not a ladies' man. More of an angst-machine. He yearns quietly, secretly (I assume). Usually stays quiet, mostly listens.

But not that night. He was talking.

Jason chimed in. Jason: in his permanent Red Sox T-shirt. "I know I want a girlfriend. And I know when I like someone. Within two minutes of meeting someone I can tell." Jason was drinking beer from a giant, mock-glass tankard. His eyes were beady. "We need more beers," he said. Got up to get them. There were four of us, but he could carry five. So he got five.

Now came my part. "I think I'm ready for a relationship, too. But, it'd have to be a pretty relaxed one," I say to a pause.

I feel like we're family, these guys and me. After weeks of Europe together, we called ourselves the Bradys. I was Marcia. I could tell them anything. Almost.

"How 'bout you, Zack?"

"I don't know. I could go either way," Zack says. "I've only really liked one girl in my life and that didn't work out, so eh . . . ," he says. "So eh" is the phrase Zack uses: a kind of verbal shrug. I've never heard someone pronounce *eh* so clearly. This reminds me of another Zack thing.

Our first day in Rome, I asked him if he felt like eating dinner. We were sitting next to each other on the careful-don't-get-grabbed-traveling-circus that is a Roman public bus. Here's how our conversation went:

Kelly: I'm pretty hungry, I can't wait to try the pizza here. Want to eat soon?

Zack (nods with no expression).

Kelly: Well . . . are you hungry?

Zack: Me? Oh . . . I don't really get hungry.

Kelly: You don't get . . . hun-gry? (Thinking if I said the word slowly enough it would spark some kind of response.)

Zack: Nah, not really. I usually just eat when other people are ready to eat.

This is what I call Zackness. Something I can't actually comprehend.

No surprise: Zack could go either way on the girlfriend front.

I was the one to bring up the doubt part. "I feel like it's just so hard to know when it's the right person, ya know? Like imagine, spending the rest of your life with one person? How does that *happen*?" I ask. This is literally the biggest mystery of my life. Even in a life where I haven't been tested for a gene that I'm 50 percent prone to.

The how-do-you-know-it's-the-right-person topic. Brought up by me. We got more beer.

I volunteered my aunt's example. "She said she just realized that she didn't want to break up this time, and that it could actually work," I said. "That's all she said."

We all pondered that. The finality of it. Then Billy brought up something his dad had told him. "The most important thing to look for in the person you settle down with is good genes," he said. One of those quotes that isn't quite a quote. Was this Billy's father talking now, or Billy?

I felt my face get hot. There was the black hole again. I felt myself getting pulled in. And then I got mad at myself for doing that and tried to stop. Then I tried to put myself in his shoes. *He didn't know*. Then I tried to think of what I should say. Should I say the whole truth, half-truth, bring in an anecdote? God, why can't I just not bring it up at all? This happened in classes, too. It happened in passing. It happened with my closest friends. It wasn't happening now, though. Now we were talking hypotheticals. Not actuals. Not me.

Billy went down another path. Talking about his father's relationship with his mother. Upset, thinking he only married her for her genes. I was already eight steps in a different direction. For me, this was a dead-end conversation.

Why is it easier to talk about genes than to talk about love?

I hadn't told them about what was in my tree.

Zack said, "I can see what he means. To a point."

Zack, who waits to eat until somebody else is hungry, weighs in.

They were nodding, the three of them. Agreeing.

I spoke up. "Well, what about artificial insemination and gene therapy? A lot can be done now," I said. Part of me feels like if I heat up any more I may evaporate into a single-forever-and-ever cloud with the other marked mutants.

Jason and Zack don't really say anything, but they are thinking.

Billy, who I always thought of as the sweetest one, shook his head. "I wouldn't really want that done. It just isn't natural," he said.

Zack agreed. "Yeah. It's definitely not ideal," he said. Always so logical.

Jason, usually the loud one, says nothing. But he nods a little.

The room got hazy. I wasn't drunk, but I was desperate. Scared. Like I was in one of my what-the-fuck-are-you-going-to-do nightmares. "I'd definitely do it if I needed to!" I said. "The idea of getting rid of diseases that your baby doesn't need to have, so that we could maybe get rid of them altogether someday, that would be amazing. Like the idea that you could right there save someone from that much suffering if they're at risk."

Pretty obvious the gene-poor princess is speaking.

They just looked at me. "I think it's so exciting," I went on, and I was excited, but at the same time I was thinking, but what about the ones that have already been born? Like my siblings? And what about me?

I could die.

"The ethics of it are really messy," they all said. Like words to a song they all knew.

I couldn't help myself. "It won't be about picking out your kids' looks," I said. "That would be way too expensive. And besides. They'll figure out how to keep it to just people selecting out dangerous diseases."

But they were in agreement, the three of them. It wasn't about getting rid of the bad. It was about starting off "better." Closer to "ideal."

"Personally, I wouldn't put myself in that position," one of the guys said.

Hard to know, after this, what I would tell the next guy I got close to. Hard to know what's ideal when there's no deal.

There was so much else I wanted to say. Maybe God didn't invite you to this particular party. But step up to it anyway and feel welcome. Because this isn't about what you have. There's a "might" in every person you meet. Hey, I like your sense of humor, I like your face, how's your karyotype? Spellcheck doesn't recognize that word. Do you?

Other things could kill you. Cancer could kill you. An accident could kill you. A volcano could erupt. An ash plume could spread from northern Europe, glittering with toxic dust, and eat your plane.

It's just time. Time will tell. Time is everyone's. Even my mom's. Even mine.

Genetics isn't about deserving. So those people can stop feeling like they are questioning "God" in the right way, saying, "He works in mysterious ways," when in fact we're staring at a way that he works in exactly proportionate probabilities. In this case, Huntington's disease, we're working with a 50-50 split. The gene that causes HD is dominant. That means that each child my mother has carries a 50 percent chance of also having this incurable thing. In our case there are four candidates: My two brothers. My little sister, Angela. And me. And please, whatever you do, don't give me the "sometimes bad things happen to good people" line. Because we are not "good people." We're just people. It's my mom, me, my sister, my brother, my other brother, my dad (but he isn't at risk here). And "we," the people we're talking about here, don't actually use the word "deserving." I'm not saying there aren't times when my mom hasn't been severely depressed (which happens to be a symptom of Huntington's disease). Like, for instance, the time she asked what she'd done to deserve this thing. But that was around the same time that she stood outside our house in the middle of the double-yellow-lined road, with my brown-haired, freckled, big-hazel-eyed 7-year-old sister watching from the bedroom window before school, and begged for someone to end her life.

Maybe my mom was thinking about "deserving," but she was also doing things like that.

We don't talk much about that time. How she acted was a symptom. It can be treated (luckily) with drugs. Also, we don't talk much about genes. When I have questions, I email premed kids I knew at school, making sure I understand my chances. My siblings' chances. If you ask the question a different way, maybe you'll get a different answer. Maybe this is my way of extending the conversation, bringing others in, making more people uncomfortable. Because I hope to do that. Extend the invitations.

Don't try to fool me, you genes-free that don't feel like you're invited. I said, Welcome! And you said, *Fuck you, girl, I ain't got the bad genes.* Well, yeah, you might not have quite as many as my mom or (maybe) as me and my siblings. I get that. When HD unzips itself in the body around age 40, it sends gene triplets into just about every cell and protein in your body. Feel it in the tips of your fingers. Feel it in every tear you cry. Sometimes I swear I feel it already. Feel it in your brain. As it shrinks. And every ability you have shrinks with it. Including your emotions. Which happens to include joy. So you don't feel invited to this party with its piñata of ailments?

You wouldn't be the only one who doesn't want to RSVP.

People don't like feeling like puppets. Especially when they can't see the strings.

Grandma Marcia, my mom's stepmom, always told me that if everyone put their problems in the middle of the table and they each got to choose which ones they could keep for themselves, everyone would end up picking the same ones they started with. So here we are.

I'm a patient for something we don't know. And I don't know if I'm a patient.

Is anyone surprised when I can't stay tied to one thing?

People say, *You are what you are.*

The cereal box says, *You are what you eat.* Girl, you're worth it! Worth what? Worth something in life. How can you say what makes a life worth living?

How prepared can you be to find out the $5,000 life you thought you'd been saving up for years, earning interest and thinking, *Wall Street better not sear my solid-money-life into something I don't want,* actually turns out to be a $3 life that no one wants?

So what is this thing about testing?

It goes one of two places, equal shares likely. It's a fairness game really. In all fairness.

It's a question of how many letters. CAG. On chromosome 4. And 4 has always been my lucky number.

If I get tested, I'll get a number: 0 or 100.

Sometimes the name of a thing is more frightening than the thing itself. Like 9/11. Maybe someday, somebody will find a word for the gaping tear that day left. Numbers aren't exactly names, though. 50 percent isn't a name, or 0 percent, or 100 percent. When I was 4, in preschool, we had to write how old we were in our books About Me. Next to our handprints. I remember wondering why the number 4 said something as important about me as my handprint did.

We made handprints everywhere. Tracing our hands. Painting our palms. Pressing them to paper. So many handprints, smooshed, over and over again in that warm paint.

But really there's just one of me. I'm just one person with too many CAG repeats. Five too many.

I don't have too many repeats. Or maybe I do. I don't know. All I have right now is my handprint.

Take it. (My hand.)

Nope. Give it back.

I need it.

I need it to pick one of those problems sitting in the middle of the table, waiting.

This is the one I drew.

It's genetics. It's time. It's life. It's love. It's a disease in my family. It's mine. Or (maybe) it isn't.

# TWO

# Driving North

Charlie Pierce

Almost thirty years ago, my father drove to a flower store to buy geraniums to plant on the family plots for Memorial Day. They were tearing up the streets of the small Massachusetts town in which he had lived since 1951. After they detoured him around the town square, he drove north and didn't stop until he got to Montpelier in Vermont. He was missing for three days.

For the next four years, he battled through the worst of the Alzheimer's disease that had been afflicting him for more than a decade, the disease that my mother had kept in their small house like a malignant secret. It nearly tore my family apart.

My father died twenty-three years ago last June. Twelve years ago, I wrote a book about it all, about the disease and my family and about the race among scientists to solve the riddle of Alzheimer's so that more families will not have to go through what mine went through. Eventually, all four of my father's siblings developed the symptoms of the disease and then died. At one point, each of them was convinced they saw their dead mother standing in front of them. One of them thought a college football team was living in his attic. The last one, my aunt, took to sending checks in odd, pin-money amounts to the Vatican. (Even more remarkably, someone there was cashing them.) Therapies had improved considerably by then. My father stopped speaking a year before he died. By 2003, my aunt was still capable of telling the same stories, over and over again. She told about the time we all walked to Mass on Christmas morning in the snow, right up until nearly the end.

My book was well received, if not necessarily well publicized. I was booked into some interesting venues. One radio host had me talk into his home answering machine for five minutes and then played the tape on the air. Another asked me one question and then suggested that, for the remaining eight minutes of our time together, we should pray. Times being what they were, I took him up on the offer, and was grateful to do so. My book passed into history; every extant new copy is now in my garage. People occasionally said nice things about it. I thought I had put the subject somewhat to rest in my life, now that I'm almost 60 years old.

It turns out that I was wrong. Alzheimer's was there most presently as my father's siblings all sickened and died, but even beyond this, it was always there, somewhere, a low-running fever of the mind. Ronald Reagan died, and there was a great burst of news about the disease that had killed him. I was still able to spot stories about breakthroughs in the science of the disease, no matter how deeply they were hidden in the newspaper, and the ones I didn't find, my wife did. New therapies, new theories of the disease's origin, relentlessly appearing, one after the other. I note them all. They mean different things to me than they do to most people.

I got older. If you gave me three things to do, I would reliably do two of them. The third was up for grabs. My typing went askew. Somehow, I kept missing the apostrophe and hitting the semicolon. For anyone else, the sudden eruption of semicolons might simply be cause to ask human resources for an ergonomic keyboard. For me, with my history, the semicolons were a different matter. Perhaps that's how I can first track my disease, through misplaced semicolons, like splinters of wood below the skin of the words—foreign bodies, suppurating infections. Now, I've entered a different place. I wouldn't say it is hypochondria. It is how things are when you are a patient with an emotional diagnosis, one whose symptoms are less than clinical but far more than imaginary.

Sometimes, before twilight, I can feel the quiver of leaves and shadows, and my wonderment is leavened with doubt about what this blissful emptiness of mind really signifies, if anything at all. Sometimes it feels like I have come to be one with everything around me. Sometimes it just feels like careering oblivion, and it's then that memories of my father's illness come back again. When I think of those days now, their fragments are like bones at the bottom of a river that shine whenever the sunlight hits them in a certain way, and then go dark again when a cloud passes. Finally, the great rushing current of the present turns them all into a blur.

The Washington University Medical School in St. Louis is a small, self-contained universe centered around Barnes-Jewish Hospital, which was created in 1992 through the merger of two major institutions in the city's history: the Jewish Hospital, which opened in 1902, and was renowned (among other things) for treating several generations of St. Louis Cardinals, and Barnes Hospital, which was founded twelve years later by Robert Barnes, a self-made millionaire philanthropist whose generosity also extended to floating a loan to a St. Louis immigrant named Adolphus Busch, who thought he might like to open a brewery. The laboratories at the medical school are just the way I remember them from the work on my book—little warrens full of postdoctoral students scurrying between the lab tables and their computers. The labs look like medieval monasteries and the folks in the lab coats look like monks, bustling through elaborate works of illumination. An entire generation of researchers has entered the field since I was working on my book.

Over that same decade, and through the efforts of people like David Holtzman and Randall Bateman, Washington University has moved to a prominent place in the Alzheimer's research community, changing how we look at the way the disease functions within its unique population. The school began its study of dementia thirty years ago. From the start, researchers there worked on the assumption that, somehow, Alzheimer's began its destructive course long before clinical symptoms became manifest, long before a person got in his car to drive to the flower store and ended up in Vermont. In theory, this meant that the disease could be diagnosed—and ultimately treated—long before its clinical symptoms had evinced themselves. "When we started looking at brains at autopsy of people in our studies who had died of some other cause," explains John Morris, the director of the Alzheimer's Disease Research Center at Washington University's medical school, "we noticed that quite a few of them had some Alzheimer's pathology in their brains without having exhibited any symptoms."

David Holtzman and Randall Bateman work in one of these laboratories. They are of that next generation of Alzheimer's researchers.[1] Since I first began to delve into the subject, Alzheimer's research has experienced explosive growth, both in the understanding of the mechanisms through which the disease works and in the pursuit of therapies to stop the progress of the disease. Once so fiercely competitive that prominent researchers would pass each other at conferences without so much as a nod, the field has become more cooperative as the available data have increased. "People have realized that this is a tough nut to crack," Holtzman says. "We really need to cooperate to get anything done."

Alzheimer's destroys the function of the brain through the gradual accumulation of a substance called beta-amyloid, a gooey protein responsible for the characteristic "plaques and tangles" that are the primary markers of the disease. Also involved in the accumulation of beta-amyloid is a protein called an apolipoprotein, APOE, a lipid-transport protein of which the body contains four varieties. One of these, APOE-4, is uniquely dysfunctional at ridding the brain of beta-amyloid, so much so that it seems to act as what the arson squad would call an "accelerant" in the course of the disease. By contrast, APOE-2 seems to do its job splendidly.

(At the time I wrote my book, a major controversy was raging over the role of beta-amyloid in the disease. Discovery of the role of APOE-4 caused many people to wonder whether the amyloid was merely a marker for the path of the disease, not its causative agent. Subsequently, that question has been settled in most minds as the latter—beta-amyloid is the causative agent.)

Holtzman began by studying people with Down syndrome, all of whom eventually develop Alzheimer's. "We knew for a long time that all people with Down syndrome get the neuropathology of Alzheimer's by age 35, but they didn't get dementia until the average age of 50," Holtzman said, when I went out to talk with him about all of this several years ago. "That tells you right there that the pathology starts way before the actual clinical manifestation.

"The brain can compensate. There's a lot of redundancy built into the brain, so you can have a lot of damage done before you actually start manifesting an abnormality that we can observe. For example, in almost every disease that we study that's like Alzheimer's—Lou Gehrig's disease, Huntington's, Parkinson's—we know that the pathology starts way before the final manifestation that we observe."

Holtzman concluded that people with Alzheimer's might well have the amyloid pathology of the disease active in their brains twelve to fifteen years before they manifest even the smallest symptom of the disease. He decided to study the level of amyloid in the cerebral spinal fluid of patients who otherwise were asymptomatic. If the level of amyloid in the CSF was high, then the amyloid was not accumulating in the brain, because CSF is one of the primary places to which it is cleared. A low level of amyloid in the CSF, however, would mean that it probably was beginning to pile up in the brain in ways distinctive to an eventual clinical manifestation of Alzheimer's. And the manifestation is, he knows, inevitable.

Holtzman talks with his hands, half-leaping from his chair. The fall before we talked, he'd driven with his father to a St. Louis Rams game. Their route took them past what once had been his father's machine shop. Alan Holtzman had been a lifelong sports fan. He coached his children's baseball teams as they were growing up, and his nephew, Ken, had gone on to throw two no-hitters for the Chicago Cubs and, later, to become a mainstay on the Oakland A's world championship teams of the early 1970s. On the way to the game, David told his father that he knew there was a parking lot not far from the old machine shop.

No, his father told him, there isn't.

"I didn't push it, but I thought, you know, that's not right, because he knows this area really well," David Holtzman recalls. "Over the next few months, I noticed that he had these issues of knowing where he was in space. It happened very slowly. He was still teaching night school at Wash U for eight years, all the way through the moderate dementia stage. His long-term memory remained good and so did his procedural memory. He could still go out with my mother to play poker. But his recent memory was very poor, and he had begun to develop apraxia—that is, the inability to carry out a learned motor activity despite still having good strength. It was as though he was forgetting how to walk." Only a few days before telling this story, David Holtzman put his father into a nursing home.

"I was interested in getting him into clinical trials," he says. "People with Alzheimer's though, they have some idea of what's going on, but their insight into their own problems is sometimes poor. He completely denied it to us. 'I don't have Alzheimer's,' he said. 'That was my sister, you know.'"

There is a certain level of energy to the way David Holtzman tells this story that makes him look, not like someone seeking a clue to the solution of an arcane scientific problem, but like a man leaping into the ring against a terrible, tangible adversary. And, listening to him talking about apraxia, I begin in my mind's eye to see semicolons where apostrophes should be, over and over again, and, still in my mind's eye, under the rushing current of the present, something begins to sparkle again.

When I think of hard decisions about the end of life, I think about the press box at Belmont Park.

It was 1989, and I was there to cover the Belmont Stakes for the newspaper I was working for at the time. My father had been in a nursing home for several months. He had fairly flourished there. They fed him well and kept him groomed. They even let his hair grow long and white until it curled behind his ears. He'd have hated that if he weren't sick, but I thought it made him look like an antebellum senator from a brand new western state, Thomas Hart Benton from Missouri, say. Or Michigan's Lewis Cass. All he needed was a frock coat, a taste for chewing tobacco, and a vote on the Missouri Compromise.

He was safe and he was cared for, and he had been neither of these back at home with my mother, so I felt comfortable going down to the races at Belmont. I had just decided to go long on a French horse named Le Voyageur, when an attendant in the press box told me there was a call for me at one of the telephones along the back wall. (This was back in that dim age of man that preceded cellphones and the ability of anyone to find you anywhere.) My wife was calling. My father had been taken ill with pneumonia. He had forgotten how to swallow, and he had inhaled some food, and now he was being rushed to the hospital. A doctor would be calling that evening to ask me what I thought should be done in the event that the pneumonia worsened. The bell rang. The race went off. Le Voyageur finished third. I made a little money back. I wrote a column on the race. Then I went back to the hotel and waited for the doctor to call. A huge thunderstorm lashed the place. The trees outside trembled and bent. I left the lights in the room off. I left the curtains open. I could see the great blank space on the wall where what light there was came through the window. The gray light was stained black in spots by hundreds of raindrops, looking like dark ash on the wall.

The call came. The doctor was punctual. He was brisk. He sounded like he was 13 years old. He wanted to know whether or not I wanted them to do anything beyond treating the pneumonia. (The proximate cause of death for Alzheimer's patients is almost always pneumonia. Through history, it's been known as the Old Man's Friend.) If they beat this one, the doctor said, there would be another one in a month, a week, a couple of days. My father was never going to remember how to swallow again. At any point, did I want them to do anything beyond treating what might be the next two, or three, or five cases of pneumonia? He was asking me if I wanted them to stop my father from dying.

No, I told him, without ever taking my eyes off the rain-mottled light on the wall. Let him go.

And that was what I had for "end-of-life" counseling prior to the death of my father. For the record, it was the second pneumonia that killed him.

I had not often thought about that night since I put it down in the book I wrote. It came to mind again a few summers back, with the initiative to reform the way this country provides health care to its citizens—I decline to call anything as chaotic and cruelly arbitrary as the current status quo a "system"—when the debate was hijacked by various thieves, poltroons, and mountebanks into a fanciful discussion of "death panels," because one proposal had within it a reimbursement through Medicare for end-of-life counseling.

The dimwit former governor of Alaska got the ball rolling. Old people showed up at public meetings screaming half-baked nonsense about socialism and Nazism and weeping about some country that was theirs and nobody else's. Younger people showed up at those meetings with guns. All of this was encouraged by a political party and a political movement whose greatest hero died of Alzheimer's in 2004. I thought then about the press box at the Belmont and about the darkened, storm-splashed hotel room, and I got angrier than I had been in years.

John Dean once famously described the metastasizing Watergate scandal as "a cancer on the presidency." This was accurate as far as it went. But now it seems to me that we are a country afflicted with a permanent kind of Alzheimer's. Our short-term memory has disappeared and our long-term memory has crumbled, and we consider history to be whatever we heard on the radio fifteen minutes ago. Our impulse control is shot, and our paranoid ravings are reaching a pitch where they embarrass the neighbors. At least partly because of the status quo in which some people showed up strapped to defend against the depredations of essential improvements, Alzheimer's is going to cost the national economy more than $200 billion a year for the foreseeable future.

And even more of the cost is hidden, carried by an underground economy made up of unpaid caregivers—elderly spouses, overwhelmed children. In 2008, the care provided by these people alone had an estimated value of $94 billion. Of the angry elderly people at those meetings, 15 percent of them over the age of 65 will slide from mild cognitive impairment into full-blown dementia. By mid-century, someone will develop Alzheimer's somewhere in this country every thirty-three seconds. Consider those figures in the light of the shift in paradigm that Holtzman and the researchers at Washington University have offered the new universe of patients. People without any symptoms at all. People who can remember everything that they're supposed to remember, clearly and precisely. Who their children are. How to speak. How to eat. How to walk. Thousands of people who are told that they are at serious risk of Alzheimer's disease or that, in fact, the processes of the disease may already have begun in their brains. Thousands and thousands of people. Do they want to know? Do they want their employers to know? Their health insurance companies?

And every one of them one day will need a doctor to ask the hard questions.

Every one of them one day will have a loved one who has to answer them.

That was what the noise was about during those angry meetings— whether or not Medicare should reimburse doctors for asking the hard questions that every one of those people will have to answer. Twenty-three years earlier, I had a press box full of sportswriters. I had the wind and the rain. Those were my "death panels." Those were the fragments that shone brightest in my memory that summer, as a country acted foolishly, dementedly. Then the light moved on, and the great current of the present rushed over them, and they were gone.

Randy Bateman crackles with energy the same way that Dave Holtzman does. He is younger, though, a farmer's kid with thinning red hair and blue eyes that make him look like Richie Cunningham come to neurology. In 2000, he was a resident, working somewhere between 80 and 110 hours a week, when he ran into Holtzman one day, doing neurology rounds at the Barnes-Jewish Hospital. "The work was such that it didn't leave me a lot of time for, ah, free thought," Bateman recalls. "Later, when I had a little more time, I asked Dave some very naive question."

Bateman wanted to know whether the accumulation of beta-amyloid in the brain of Alzheimer's patients was a matter of overproduction or a failure of the body's ability to clear the substance out. He uses the analogy of a sink. "You have a faucet, okay?" he explains. "And you have a sink and a drain. If, at any point in time, you go up and measure how deep the water is in the sink. That's the equivalent of going up and measuring how much amyloid-beta is in the cerebral spinal fluid and how much is in the brain.

"What it doesn't tell you is how much water is coming into the sink from the faucet and how much is being drained away. You can't know that unless you measure the faucet or the drain directly, or if you throw some dye into the water. What'll happen then is that all the water will turn blue in a period of time. If you stop turning the water blue, it will get cleared away. That's a function of how fast the faucet runs and how fast the water's draining away. You need to watch over it for some time, and you need that dye." So what was causing this excess beta-amyloid?

Holtzman told him that this was a very good question, because nobody had been able to figure it out yet. There was no way to "dye" the amyloid in the brain because there was no marker substance that could pass through the blood-brain barrier. Bateman, however, had an idea, one that had been spawned in his work with a marker substance called carbon-13, a nonradioactive form of carbon present in all living things. He had worked with carbon-13 before, both as an undergraduate and in medical school, marking proteins in muscles and in the bloodstream. "He said, 'What we have to do is put in a lumbar catheter,'" Holtzman says.

"I told him that as long as the C-13 gets across the barrier, in theory, this should work. He said, 'I think I can do this.'"

There was one drawback. They needed a human subject. Bateman volunteered himself. Then, there were a number of issues, most of them bureaucratic. "I wrote up the protocol to seek approval," Bateman recalls. "First, they rejected it. They told me I had to write up a consent form. I disagreed with the initial recommendation. Why would I have to write up a consent form? Why would I have to write it, then read it, and then sign it myself? It made no sense to me at all. Of course, I understand what I'm doing. I'm the one designing the study. They said, those are the rules. So I wrote a four-page consent form. I explained all the risks to myself. I explained what the research benefits were. I had someone witness it. Someone saw that I had signed the consent form that I'd written myself. Then, I filed it."

The Human Studies Review Committee told Bateman that he could not be the investigator, the subject, and the admitting physician simultaneously. So he got Holtzman to admit him and to administer the carbon-13 for a day. Then Holtzman drew off some of Bateman's spinal fluid for analysis. The carbon-13 had marked the beta-amyloid perfectly. They could now measure with confidence the rate at which it was building up in the body and the rate at which it was being dispersed. The nature of the disease process was open to them, even at its earliest stages. Long before people found themselves driving north and getting lost.

"People thought that beta-amyloid was this slow-moving protein," Bateman says. "Probably the most important thing that came out of the research is that it almost completely turned that thought over. Once we did the labeling, and measured how much was being labeled over time, this protein had an incredibly rapid rate of turnover. It wasn't this slow-moving insidious kind of protein. It was being produced and cleared at a rapid rate. Half of it's being made and cleared in about eight or nine hours. In the average workday, you're clearing half of your stuff out.

"Every human being has it. The basic question is, are people making too much of this stuff and that's why it's building up, or they're not clearing it away as well. More and more scientists are coming to the realization that, by the time people have the symptoms of Alzheimer's disease, having trouble with their checkbook or whatever, a lot of damage has already taken place. The plaques have been there a long time. The tau-tangles have been there a long time. Those neurons are dead. The brain is one of our best organs for compensation. But very few organs can compensate if they lose 60 or 70 percent of their cells. If you lost 70 percent of the cells in your heart, you couldn't walk down that hallway."

The application of these techniques is obvious. If the new paradigm is to treat Alzheimer's before it becomes highly symptomatic, which makes sense—after all, it's better to treat your hypertension before you have a stroke—then the carbon-13 marking procedure is valuable, not only in identifying individuals at risk, but also in judging the efficacies of the new therapies devised to fit the new treatment paradigm. "Can we do something to prevent them from getting the disease in the first place?" Randy Bateman asks. "That's where things are heading. You know, none of my patients come in and say, doctor, I'd like you to freeze me in this mild Alzheimer's disease. What they really want to do is to go back to normal. And the odds there are lower because it's hard to build a new brain."

I took the MetroLink train from the medical school campus out to the airport in St. Louis. People came and went at every stop. I didn't notice them board the train and I didn't notice them leave. I was lost in thought. I was lost.

What do we do with a new universe of patients that subsumes the old one, the clinical one, the one that leaves people driving their cars somewhere they didn't plan to go? What vocabulary do we devise to tell people that they have the beginnings of Alzheimer's disease a decade before anyone will ever notice, including the patients themselves? This is to ask otherwise healthy people to define themselves as patients in a world not always comfortable with that term, a world of denial and doubt and "preexisting conditions." Tell someone that he or she has the disease, but no symptoms, and life itself becomes a preexisting condition.

In the summer of 2004, after my aunt died, we arranged to have samples of her genetic material, and that of my father, which had been stored since his death fifteen years earlier, sent to the Toronto laboratory of Dr. Peter St. George Hyslop, a towering figure in the history of

Alzheimer's genetics. That fall, Hyslop reported back that he could find no trace of the mutations known to cause Alzheimer's in either my aunt or my father. That was something of a relief. Still, there were four siblings, and they all had the disease, and I was the oldest one in the family left.

On the train to the airport, it struck me that my life had begun to take on many of the aspects of a patient's life right around the time my father disappeared. It struck me that I was not a very good patient. They had found no genetic marker for the disease in my aunt, but Holtzman had told me that afternoon about a study being done at the center that involved families like mine—families with several Alzheimer's patients, who had no apparent genetic link to the disease.

"Polygenic," he called them. The study was being led by a formidable British woman named Allison Goate, about whom I had written in the book.

So it was possible to have a familial link that wasn't directly genetic. I should be taking my antioxidants. I should adjust my diet. I should do the things that a patient would do. Still, the sense that I am somehow a patient has never encompassed these simple things. As a patient, I'm more spectator than participant.

I look at many things now from the aspect of what has become our family disease. I read the newspaper in a different way, spotting the relevant news no matter how deeply buried, approaching issues through the prism of this disease. My anger at the way people had tried to hijack health care reform undoubtedly had something to do with this: what the cost of treating this disease would be going forward, the human toll it would take upon its victims and their families, and the inevitable sterility and inutility of the debate.

This is the way patients think. This is the way I think now. If Holtzman and Bateman are right, then the disease began in me a short while after it became manifest in my father, and that was quite a while ago.

Am I my own death panel?

Am I having end-of-life conversations now?

I was thinking about these questions when the train stopped at the airport station. I was thinking so deeply that it took me a second to realize where I was and what I was supposed to do. I didn't remember any of the other stations. The other passengers around me were strangers. I didn't remember any of them getting on the train. I got up and walked onto the platform. I had no answer to my own question, and I had absolutely no idea how long I had been wrestling with it, lost in what felt like fierce concentration.

Might have been fifteen minutes. Might;ve been an hour.

# THREE

# Collateral Damage

Kate Preskenis

·············································································································

I've been afraid of Alzheimer's disease for twenty-five of my thirty-seven years. This disease, a huge, looming presence, deconstructs and decimates the lives of those I love.[1] Five of my family members have died from Alzheimer's, not counting the death of my father, who died, emotionally exhausted, as my mother's caregiver.

The first time I remember hearing Mom talk about Grandma's experience with Alzheimer's, I was in eighth grade. Mom told me then that the AD in our family might be genetic. I knew Grandma loved to sing, relax at the beach, and spend time with her sisters. I knew she savored butterscotch candies. But I didn't know that at age 39, she'd forgotten to pick her children up from school and misplaced Grandpa's paychecks. Nor did I know she endured shock treatments for what her doctors thought was postpartum depression after giving birth to her tenth child. She died from early-onset Alzheimer's disease when I was 3, after a long, fifteen-year battle with the disease.

It disturbed Mom, remembering Grandma at the end of her life, kept alive by feeding tubes and unable to recognize her own kids. Mom told me that if she became mentally incapacitated and couldn't feed herself, she didn't want to be fed.

I hated listening to Mom talk about the possibility of dying. Besides being my mother, she was competent and skilled at nearly everything that mattered to me. Mom did the carpentry, plumbing, sanding, painting, and electrical work to remodel our home. She cut our hair, made creative birthday cakes, baked bread, sewed skirts for my chorus performances, raked leaves, shoveled snow, and did all this so effortlessly that my siblings termed it "Mo Gear." Mom's nickname was Mo and she could multitask faster and more efficiently than anyone we knew. After her symptoms began, that changed.

As many as one in eight Americans over age 65 currently have Alzheimer's. In 2012, the Alzheimer's Association estimated that 5.4 million people in the United States have this disease, which—combined with other dementias—will cost us approximately $200 billion a year. Roughly 5 percent of AD patients (two hundred thousand) have young- or early-onset Alzheimer's, meaning symptoms start before the age of 65. One percent of Alzheimer's cases are autosomal dominant Alzheimer's disease, ADAD, also known as early-onset familial Alzheimer's disease, EOFAD. My family falls into this category. If an individual has one of the known genetic mutations (PS1, PS2, or APP), he or she will almost certainly get the disease, often before the age of 50.[2]

In Mom, AD appeared slowly; it was almost unnoticeable. It wasn't defined by a list of clinical symptoms. I had a sixth sense of something amiss, an innate knowledge that had no real criteria. This had to be lived with to be recognized; I had to know Mom intimately to detect the change. This could never be fully transferred from my experience, into words, to a research doctor. The early symptoms were too elusive.

When I was in high school, Mom stopped making dinner. This was a stark contrast from the years I was in elementary school, when my father was in seminary, training to become a minister. We had little money, so Mom found creative ways to fill our bellies. She transformed unlabeled, dented cans of soup into casseroles. She picked up day-old donuts from a bakery or salvaged wrapped mini candy bars that had been cast aside for pig food. Mom's homemade bread was a staple back then.

If she didn't want to cook after all these years, it seemed justified.

The odd changes continued.

On family outings when I was a young girl, Mom thought we should never let a little rain "ruin our parade." If we had plans for swimming, we went, regardless of the weather. "It's just a little water. Who cares if it's raining if you're already wet?" When I was in junior high, she'd attended most of my soccer games, home or away. By the time I was a senior, Mom stopped coming to my home soccer games if it was cold and wet outside.

When I was learning to drive, Mom had taught me how to steer out of a skid on black ice. She insisted I practice in winter weather so I'd know how the car reacts. But when I was in college, she started having accidents in snow and ice, ending up in a ditch multiple times.

Were these shifts due to personality changes or a reordering of Mom's priorities? Mom no longer traveled the five hours to visit her younger sister, Fran, who was her best friend in my estimation. Familiarity of location took the place of connection. The tragedy: Fran was also beginning to show symptoms of Alzheimer's.

Mom bought Butter Rum Lifesavers by the case. She'd put them in sandwich baggies for easy access, then, like a chain-smoker, pop one after another in her mouth. I hate that I like buttery candies too. I purposefully don't partake, but I salivate whenever I see the golden candy.

Mom adopted a teenage attitude: searching, teasing, wanting freedom. Was this a classic mid-life crisis? Or an about-turn of a sailboat returning to the same direction from which it had come, an unlearning, a reversal, a decay?

Her eerie blank look, the look of Alzheimer's, stole longer lapses of time.

When I was 26, my family's main question, *Why are we getting Alzheimer's?*, was answered. Forty members of my extended family met at Brigham and Women's Hospital in Boston. We sat in a semicircle around two research doctors, Dr. Albert Green and Dr. Eddie Watson. Researchers from the National Institutes of Health had found the mutated gene that causes Alzheimer's disease in our family. The doctors had called this meeting to explain what this meant for my family: the potential impact of this discovery, the opportunity to learn our genetic status, and the importance of our continued participation in research.

Several important people were missing. Mom wasn't there—she was in a care facility, heavily medicated to subdue her angry outbursts and brute strength. The last time she'd participated in research at the NIH, she had given blood and undergone an MRI. Her steep mental decline precluded participation in cognitive testing or home skills. She needed one-on-one supervision, but even that didn't prevent her anger when she kicked a garbage can down the hall. Eventually, after her death, Mom's brain would become her final contribution to research. Fran was also absent from this meeting; she was in a locked care facility, restricted to a broad walker that she couldn't reach beyond. She moved throughout the building, but the walker kept her fist out of reach of the other residents. Like Mom's, Fran's aggression was also the result of confusion combined with an able, youthful body.

At the meeting, a buzz hung in the air like static electricity, sparking to life with any movement. I was glued to the chair. *What if I missed something important?* It all seemed important.

"What was special about your family was upfront right from the beginning," Dr. Green said. " 'We are concerned about ourselves, but we are also concerned about our children,' " he recalled someone in my family saying.

I had attended a family meeting with Mom and Dad when I was in college. That was back when Dr. Green had first talked with us about Alzheimer's and asked if members of Mom's generation would give samples of their blood. I was disappointed they didn't want my blood, too. However, the research was still in the preliminary stage, as the doctors were trying to learn whether our family's Alzheimer's was genetic.

Dr. Green continued, "This wasn't the typical one-to-one doctor-patient relationship. It became clear your family members aren't just interested in themselves or their children, but have a broader goal: research, because that will benefit the wider community. Nowhere is this better exemplified than in the Memory Ride that some of you created, which has been an extraordinary gift to the Alzheimer's community."

I glanced at my uncles John and Eryc, who began the Memory Ride. Tired of feeling helpless in the face of this disease, they organized a bike ride to raise awareness and money specifically for Alzheimer's research.

As Dr. Green spoke, Butch, another uncle active in the ride, operated a video camera. In his late forties, Butch had recently developed early symptoms of Alzheimer's. Eerily, the disease continued to creep through our family.

"Right from the beginning we were looking for a gene . . . The gene that can cause Alzheimer's disease in your family has one letter in it that contains a mistake, which, when inherited, leads to Alzheimer's. When you extract DNA from someone's blood, you find this deviation," Dr. Green explained.

"Even though we can't say when, if you have the mutation, you'll get this disease, unless something else strikes first. Nothing is a hundred percent in medicine, but I would say there's a 99 percent probability you'll get the disease if you have the mutated gene. The probability is very, very high." Dr. Green adjusted his glasses. "It is a serious, serious mutation. This conversation today is a big deal."

He took a sip of water and cleared his throat. "You each have another decision to make, in the context of the most significant things people do, like getting married and having children and even entering into relationships. Because when you know there's a possibility of the mutation, it's in your power to decide if you want to be tested and who you will tell."

My father had died two years earlier of a massive heart attack. Before he died, I'd never heard of caregiver death. When I learned it was common, I was angry that this might have been prevented by education and intervention. My parents were treating Mom's illness with both traditional medicine and alternative therapies. This was so expensive that my dad was on the verge of claiming bankruptcy and had taken a second job in addition to full-time work. I blamed Alzheimer's for killing Dad. He had become one of those statistics: "collateral damage."

I couldn't imagine inflicting a potential husband with caregiver's stress if I were to succumb to this terrible disease.

I pulled myself back to the discussion.

"We are now entering the waters of genetic counseling," Dr. Green said. "Some people think of planning in a financial sense, providing for their children. But I'm going to be very frank with everyone here: for other people, when they talk about planning, they're actually talking about suicide. There are studies—one occurred very recently in Sweden. Researchers found a family with a mutation and after they were informed, a number of people in their family did take their own lives. Because this is so serious, we need to know the guidelines for genetic counseling."

I gulped, seeing the grim, ceramic faces that were also looking at me and at each other.

Dr. Green went on: "Remember, no one is being forced to contribute to this research. It is your individual decision. It is not your family alone that shoulders this burden. There are other families out there already participating in research."

Seated by the window in the conference room, a cousin tried to understand our options. "The first scenario is that we line up, get blood drawn, and we never see you guys for the rest of our life. It helps you in research some, or maybe it doesn't, but I will never come back here again and I am done with it. That's one—"

An uncle interrupted in a sing-song tone, "That is one, but that is not the one you are going to do."

Laughter followed.

Another uncle's deep voice cut through the laughter, trying to keep us focused. "The second scenario is you give blood, you agree to do research, and you come back every year, not here, but to NIH. You go through the lumbar punctures, the blood work, the cognitive testing, just for the research. Then there is a third scenario, where you do all of that and you want to know, you get genetic testing and find out your status."

A relative contributed, "And there's a fourth, which is you do all the research, you don't want to know, but you then go for the genetic counseling because somebody else might want to know and their status affects you and you'll still have to deal with it emotionally."

A person from the older generation who rarely talks about this subject said, "Then there's a fifth." His strong arm flinched as he talked. "The fifth is if your kids have it," he shook his finger at the first cousin who'd talked about scenario number one, "then there is an opportunity for your kids to be healed of this disease, and if you don't go through it, someone else over here is going to go through it," he pointed across the room, "and you are going to owe them something at some point in your life, because they spent their time and their energy for the main goal: they want to heal your kids and their kids and the wider world from the presence of Alzheimer's disease. So there is a major goal involved and it is healing the disease—it is killing the disease before it keeps going so we don't have to be here again. I will probably be long gone, but the positive end is ending the disease."

Gretchen Taylor, the NIH nurse who'd helped with Mom last year at the Memory Ride, chimed in. "Be careful you don't draw a wedge between you and your family members. Respect each other's decisions, even if they aren't the same as yours."

Dr. Watson added, "I vividly recall the sounds of going to the nursing home to visit my great-great-grandmother." He pointed to his own chest. "I still get the heebie-jeebies when I go into nursing homes. If you were a child when you first saw this illness, you'll have a different perception of this than someone who's older." He motioned to the younger and older

Noonan siblings, who were nearly twenty years apart. "This is well documented in other illnesses: each of you has a different take on it, and that's how you cut each other some slack. Because what each of you sees is based on the age of your eyes, your brain, where you are in your life."

He looked around the table. "Many of you by the time you were 10 had changed a lot of diapers on younger siblings."

I was in high school when Mom stopped cooking and doing laundry. My cousins were still kids when Fran got sick, and they had to pitch in and help.

Dr. Watson paused. "Now, the whole prospect of being able to get a genetic test to tell you whether you're going to go down that same path and be a burden, be aggressive, embarrass your family—these are the things it conjures up. It doesn't have to be true, but that may be how you've experienced it. You have to take that into consideration, to see if you really want to know. It is a complex process, to say the least."

Silence fell across the room. The white tile floor was squared off with silver lines. I saw my water bottle next to my chair and reached for it.

Fifteen minutes later, Gretchen Taylor, the nurse from NIH, secured a rubber strap around my upper arm and told me to flex my hand into a fist. I was glad they finally wanted my blood, but I felt dizzy and overwhelmed. Quickly, she inserted the needle and blood poured into the attached vial.

The nurse rechecked the small plastic tubes to be sure they were assigned to me and then put a piece of cotton and a Band-Aid on the point where she'd extracted the needle from my flesh.

I felt simultaneously obligated and desperate to participate in genetic studies, both for me and for my family. It was our only hope. The blood draw wasn't a problem for me, but a spinal tap? A lumbar puncture that could leave me paralyzed?

Did I want to know my results? Could I handle the results? What if others found out theirs?

Two years after our meeting in Boston, my siblings and I met at the NIH. We held the heavy gray door open for each other as we filed inside. I was 28. The sterile smell of cleaning products entered my nostrils, forcing the robust outdoor air out of my lungs. CLICK. The door shut, locking us inside. Voluntarily admitting myself to contribute to the National Institutes of Health clinical study, I was comforted only slightly by the presence of my siblings and my feeble hope this might lead to a cure or treatment for Alzheimer's.

We separated, each of us undergoing intake tasks of blood testing, EKGs, surveys of medicine and vitamin lists, and medical histories. I was asked to provide a urine sample. Peeing into a plastic container, I heard a commotion in my bedroom. I was in a bathroom that didn't have a lock, so I rushed to yank up my black cotton yoga pants.

I turned the metal knob on the door to enter the room.

A middle-aged woman took two steps toward me and said, "Hello." On the far side, near my luggage, my intake nurse was focused on me, watching this interaction and my blank expression.

"Hi, what's your name?" I asked.

Drawing back, the woman looked insulted. "Susan," she said. Her expression suggested I should have known who she was.

I didn't remember her and groped for words, trying to understand. "My name is Kate."

"Yes, yes, I know, we met last year."

She continued to stare at me, clearly expecting something, although I wasn't sure what. My mouth went dry—I hated not remembering. I bit my lower lip. She must have been a nurse of mine during my first visit to the NIH after Mom died. Or did she attend the Memory Ride the previous year with other NIH staff? I scanned my memory. Nothing there. I tried to convince myself. Still nothing.

The confusion of not remembering Susan and my inability to read the social cues rattled me. I turned to leave, sneaking out from under the scrutiny of her expectant eyes.

In the hallway, another nurse, seeing my bewilderment, said, "So you've been talking with Susan?"

"Uh, yeah," I said hesitantly, rubbing my chin. *Who is Susan?* I wondered. *Why can't I remember her? This is an early sign of Alzheimer's.*

I searched the nurse's face for answers, and she smiled knowingly.

Seeing her smile, I asked timidly, "Is Susan my roommate?"

"Yes," she called over her shoulder, continuing down the hall.

Suddenly, it all came into focus: *If Susan is my roommate, it is likely she has Alzheimer's symptoms. I don't remember her because we never met!* Susan had tricked me into thinking I was the forgetful one.

Relief swept through my body as I leaned back against the wall, arms hanging limp.

I used to be skilled at knowing when Mom was utilizing her survival tactics to navigate the world. Ironically, in this locked ward, I wasn't prepared to interact with someone else with AD, and I'd been duped by the same basic coping tactic my mother had used.

As I listened from the hallway, I heard the intake nurse inside the room ask Susan, "How old are you?"

Susan responded, "Why keep track anymore?" As if verbatim, that was Mom's exact delivery, tone, tempo, and punch line.

At 29 I was living in Marin, California, with my boyfriend, Andy. He worked and I wrote—he was giving me the gift of concentrated time to work on my memoir. My favorite way to procrastinate was by scanning Craigslist, looking for part-time jobs. Ads for egg donors made quick money look easy. It was tempting—I could earn extra cash, it wouldn't take much time, I'd be helping another couple conceive. I bookmarked the ads: *prefer fair skin, blue eyes, college-educated, athletic.*

For a few weeks, I actually considered it. I was getting ready to discuss it with Andy. Then, thinking of the Alzheimer's gene, I had to laugh at the absurdity of this fast-cash plan. If I didn't want to have children without knowing my genetic status, how could I possibly sell my eggs to another couple?

My next thought: If I found out my status and I had the gene, maybe I would get pregnant anyway, terminate the pregnancy, and donate the fetus to science. How many eggs did I have left? Could I donate them all? Would the stem cells help Alzheimer's research? Could the eggs remaining in my ovaries save members of my family who were already living? How many embryos could I conceive before my tender heart or a research ethics board intervened?

A few months later, I took another break from writing to spend time with my cousin, Jessica, who had stopped in for a visit during a spring-break college road trip. As I rounded the corner into our wood-paneled living room, I saw her sitting with her friend on our bark-brown love seat. When she saw me, she stopped talking. I stutter-stepped, thinking I'd interrupted, but she was looking intently at me. Wonder filled her face—her mouth dropped slightly, as if seeing something for the first time.

She exclaimed, "Katie, you look like your mom!"

Uncontrollably, I shuddered as three lightning bolts struck me.

My mind jumped to the genetic links and the fear that I will end up like Mom did. I had a sudden memory of her, drugged and hunched over, drooling and half asleep over a tray of food. Mom's hair was nappy and dirty, and she had food on her face and pink medicine between her teeth.

With the next bolt, I shivered, remembering Mom's anger and hearing her yell, "That's bullshit." Her fist was clenched in defiance.

The final memory was a zap from November 23, three years earlier. Dried, tight tear lines cracked on my face. I was exhausted, lying limp next to Mom on her hospice bed, cradling her still-warm dead body.

It was an instant after Jessica spoke and I was only three steps into the living room. Jess saw my shudder and quickly asked, "Is that bad, to look like your mom?" Then she added, "I mean when she was young, of course." Her wide blue eyes waited for an answer, trying to be sure I was not insulted. Her hair, a silky chestnut-brown with a reddish hue, was the same length as mine and skimmed her lower shoulder blades. She was biting her fingernails.

I cringed, knowing she was proud of Fran, her recently deceased mom, and that she wanted to be like her—a smart, powerful, outspoken, devout Christian. Her mom and my mom were the first of their generation to get Alzheimer's disease. I didn't want to influence Jess with my anti-Alzheimer's obsession, so I stammered, "Uh, no, I guess not."

I continued into the kitchen. Once out of her sight, I grabbed the edge of the old wood cabinets with copper handles, to steady myself, and closed my eyes. *How did I let my true feelings show?* Chatter started again in the living room. It was deafened by the blood pulsating in my eardrums and the words *You look like your mom!* reverberating inside my skull.

Several years later, Andy lifted the garage door of a house I was renting. I smelled a whiff of mildew. Halfway back on the left was my pile. Andy was helping me move boxes of books to reach the one with my high school yearbook. We had looked through his yearbooks, but mine had always been in storage. After slicing the tape with his pocket knife, we opened up the box and removed my yearbook.

I looked innocent, happy, vibrant in those pictures. Naively innocent. At that time, Mom was beginning to forget. In one picture, black and gold paint covered my face and polka-dotted my legs. I was on the shoulders of another student also voted the most school-spirited. I was holding the school's spirit stick over my head.

I opened the next box, marked "Keepsakes," knowing the contents. I saw yellow fleece material on top and slowly tucked my hands around the middle so the bundle inside didn't fall out. Flipping the bundle over, I saw dark plastic lashes open to reveal blue eyes. *My baby-doll.* Tears gushed out of my eyes. Holding the life-like body to my chest, I could feel her familiar molded legs, toes, fingers. Full sobs rocked my body. *Do I want to have a baby?* I touched her mostly bald head, the result of my frequent practice of cutting her hair, which my sister would try to fix, and then I would cut again until only stubble remained.

Wet-faced, throat tight, shaking, I looked up at Andy, shrugging my shoulders. "I don't know why I'm crying."

He wrapped his arms around me. "You don't have to know."

Gently, I laid the doll back in the box and wiped my eyes with the cuff of my sweatshirt.

Now, at 37, there are times when my peaceful sleep is held hostage by a nightmare. In one particularly vivid dream, waxy, cold, jaundiced skin is sliced open in the center, the beginning of an autopsy. I want to take the knife, but feel odd. This is the body of one of my relatives. This brain will join the other brains of family members in a research laboratory. Part of me wants to watch the autopsy, but another part is queasy—I knew who this person was before Alzheimer's disease took away the memories.

This person had the gene, but what triggered the disease? Why was the onset later than for other family members? What determines when it begins? Vitamins, exercise, happiness? Stress, fear, addictive behaviors? Each year, researchers find out more and more about AD, but it's not enough for those of us who are waiting.

In my dream, I look at the autopsy table and panic seizes me. I grab the pathologist's arm and implore, "We need to save the whole body for research, not just the brain. Do the regular autopsy, but save everything."

I wonder, *How could we be so foolish? We couldn't have known, but if only we had kept the whole bodies, we could have been more help to research*. Researchers are currently studying what turns genes off and on, the so-called activation mechanisms. Are preserved brains useless without their bodies, to see if there was a liver malfunction that triggered the "on" position of the AD cells, the plaques, the amyloid? Isn't saving only the brains like performing cognitive testing without the supporting biological tests of blood draws, spinal taps, and brain scans?

I continue, "You have to tag the body, but also, it needs to be branded or tattooed with the person's name, the disease, the gene that caused it, with the date of birth, date of onset, and date of death." I touch the leg on the table. It feels rubbery through my gloves, but I'm imagining the vital information inscribed on it. Maybe Alzheimer's is like pneumonia, something weakens the immune system and then it takes over. If we are proactive, our bodies can be studied by future scientists, allowing them to see the disease as a whole system.

My heart is pumping hard, my breathing shallow. Cautiously, I open my eyes, the autopsy is gone, I'm in my apartment. It was a dream. Moonlight from the skylight throws shadows on the walls. My hands are clenched—grasping—nothing. The chance for a cure has evaded me again. I follow the moon shadows to the kitchen, where I fill a glass with water.

The emotion of my dream is still palpable. I'm tormented at the loss of valuable research material and the fact that the "significant advances" in treatments that were hoped for when Fran and my mom began having symptoms, twenty years ago, have not yet materialized in a form that will help us. In my kitchen, these feelings combine with real life. The nightmare isn't over now that I am awake. It is worse. I'm terrified and despondent that we can't save ourselves and are still desperately waiting on researchers to find a cure.

My generation is next, and I don't know how we will fare.

*When do I learn my genetic status? Do I wait till there's a cure, or when finding out will change my life course?*

For now, the sliver of hope that I don't have the dreaded gene is more important than finding out with certainty. To find out with certainty grants total hope or seals all hope away.

# FOUR

# In Samarra

Amy Boesky

..................................................................................

There was a merchant in Bagdad who sent his servant to
market to buy provisions and in a little while the servant
came back, white and trembling, and said, *Master, just now*
*when I was in the marketplace I was jostled . . . and when I*
*turned . . . Death . . . looked at me and made a threatening*
*gesture, now, lend me your horse, and I will ride away from*
*this city and avoid my fate. I will go to Samarra and there*
*Death will not find me.* The merchant lent him his horse,
and the servant mounted it, and he dug his spurs in its
flanks and as fast as the horse could gallop he went. Then
the merchant went down to the marketplace and he saw
[Death] standing in the crowd and [asked him], *Why did*
*you make a threatening gesture to my servant when you*
*saw him this morning? That was not a threatening gesture,*
[Death] said, *it was only a start of surprise. I was astonished*
*to see him in Bagdad, for I had an appointment with him*
*tonight in Samarra.*

> "The Appointment in Samarra,"
> retold by W. SOMERSET MAUGHAM
> (1933; italics mine)

My mother loved the story of the servant who fled from Baghdad
to escape Death, only to meet him again in Samarra. For her, it
demonstrated the power of fate, as well as the urge to resist it. That story
ended up having uncanny applicability for my family. My mother tried
hard to avoid the ovarian cancer that had claimed her mother and aunt.

And she *did* avoid it. With my father's urging, she had a complete hysterectomy in the 1980s, soon after her cousin, Gail, was diagnosed with stage 4 disease. But just before the BRCA gene was discovered, my mother found a malignant lump in her breast, and as the first tests for the mutation were hitting top cancer hospitals in the early 1990s, she died from breast cancer, something none of us had ever worried about. Her own Samarra, waiting for her, despite all her efforts to evade it.

As I look back now on my mother's story and its impact on my sisters and me, I'm struck by how much genetics has become part of our discussion over the past generation. Tests for BRCA1 and 2 are widely available today (despite ongoing disputes about the patents held on these genes and the consequences for both the price and availability of testing). Today, many hospitals and clinics offer trained genetic counselors, ready to help individuals and families understand the range of therapeutic options. Rapidly expanding information may help guide women in planning the best age for surgical or other forms of intervention, if they choose that option. The surgeries themselves have been refined, and—importantly—there's now a great deal of shared information available through advocacy groups and patient networks to support women as they make these decisions. New trials are underway using prophylactic agents to reduce the risk of cancers in young, BRCA-positive women before or even instead of surgery. Organizations such as FORCE have worked to mitigate the isolation and confusion many people experience when they learn they carry the BRCA1 or 2 mutation, which can increase the risk of breast and ovarian cancers up to 90 and 50 percent, respectively, over a woman's lifetime. We've made important strides, not least in the openness with which many women face the news of this mutation and explore its consequences.

In other ways, though, little has changed.

It's easy to believe that everything we know now is true. The past looks silly, in retrospect. We forget how quickly our own data and decisions will become history, replaced by new data, new truths. It's easy to see the medical options available to us in the "now" of the present as the right ones and to lambast the choices and decisions of the past.

Yet I'm continually struck by how little we still know. What we mean by *knowledge* is, in itself, so limited, so laden.

Francis Collins has suggested that the desire to learn about a genetic mutation can be represented by an algebraic formula. The greater the chance for a remedying action, the greater the desire to know. Put simply, the more I can do to intervene, the greater my desire to learn what it is I have.

Yet, even after a woman positive for BRCA1 or BRCA2 has her ovaries and breasts removed, she still "has" the mutation. She continues to worry. She may remain at heightened risk for other cancers, few of which can be found through effective early screening. Pancreatic cancer. Cancers of the colon, stomach, eye, or skin. Surgery may leave her with physical and emotional scars. Surgical menopause at a young age may lead to a host of issues, among them, an increased risk of heart disease and osteoporosis. Then, there is the psychological freight. She may miss feeling "normal"—shucking off her shirt in the dressing room of a gym, getting undressed in front of her daughters.

Daughters. That's the other thing. What does she tell them, and when?

We shape our stories as they shape us—backward and forward across time. My role with my daughters has always been one of reassurance, of tempering fear. One step at a time, I used to tell them, leading them outward into the world. But then mine has also been the role of protector, pulling them back to safety. As I look forward, through them, past them, it's harder to see the contours of their paths.

I grew up inside of stories. My mother was a historian long before she went back for a master's and began teaching AP European History. For her, history was primarily about personality—her interest lay not in large cultural sweeps, but in the ambitions and conflicts of individuals. Cleopatra. Elizabeth I. Robespierre. Napoleon. Trotsky. History was shaped by their passions, their disputes. With family history, it was the same. The broader details of her family's past were conveyed only through sketches: a border village between Russia and Poland in 1905; sepia-toned images of poverty and striving. In stronger strokes, she drew for us the character of her grandfather Meyer, a boy of 14, drafted into the Tsar's army, urged by his family to leave under cover of night, taking with him nothing but a few coins knotted in his sock. Enough to buy steerage to Marseilles, where he could board a second ship for America, to begin his real life. The Chicago life of a self-made man, a fate-maker.

My mother didn't know her grandfather well. She learned most of what she knew about him from her mother, Sylvia, Meyer's adored older daughter. Most likely, Meyer brought more with him from Eastern Europe than that knot of coins and steely self-determinism. Likely, he also brought the "deleterious mutation" on the BRCA gene that got passed down, as bad luck would have it, to both his daughters. The mutation that copied itself over and over in our family, shaping us in ways we understand and ways we (likely) never will.

But then, Meyer gave everything to his daughters: those moody, fair-haired girls, Sylvia and Florence, with their new American names and nicknames: Sis and Pody. Their assimilated tastes for jazz, lipstick, Edna St. Vincent Millay. Their snazzy slang that eluded him. Their bright, ironic eyes. They bloomed in the shadow of their mother, Bea, an irritable, despotic woman who lived to be "about a hundred," in my mother's estimation, whereas Meyer died young of a stroke. My mother's story of his death smacked of magical realism. Meyer managed, she told us—despite being bedridden—to haul himself to his bedroom window and hurl himself out. She recounted this with such admiration that we never questioned her. How could a bedridden stroke victim throw himself out the window? And if he could manage that, why not stick around and try for rehabilitation? My mother skirted detail. It was Meyer's character she admired. He spared them his suffering, she explained. Meyer exemplified the triumph of human agency over the vicissitudes of things. *He cheated fate.*

Well, not really, we countered, tentatively. *Since he ended up dead.*

That isn't the point, my mother said. It's *how* he died.

My mother believed in control. Deeply, with the devotion some people reserve for religion. But she also believed in fate. Things are written, and you can't un-write them. Look what happened to Sylvia.

Sylvia was my mother's world. My mother's father, Jerry, was an angry man, a drinker, careless with money, a shirker of responsibilities, large and small. Sylvia divorced Jerry back when divorce was taboo, and that single act, like Meyer's alleged suicide, earned Sylvia my mother's perpetual admiration. Just like that, you could remake your life, set yourself free! My mother and Sylvia lived more like roommates than traditional mother and daughter, in ever-shrinking apartments, clinging to shreds of

Meyer's inheritance and remnants of their earlier, finer lifestyle—silk stockings, French perfume. Sylvia's heart broke when my mother left for the University of Michigan at 18. In some versions of the Sylvia story, my sisters and I enfolded that heartbreak into what followed. In others, grief merely set the mood. We knew this: my mother came home the summer after her freshman year to learn Sylvia had inoperable ovarian cancer. Her doctors gave her three months to live. She died in September, at 43, just after my mother's nineteenth birthday.

Everything else about my mother came back to Sylvia's death, like filings flying to a magnet. Her grief determined her. Because of Sylvia, my mother took a semester off school. Married my father, even though she was only a sophomore. Finished college in married student housing. Began, then left, a doctoral program in clinical psychology. Threw herself into becoming a mother, having three daughters before she was 30. Devoted herself to my sisters and me. We were the future: she kept her eyes fixed on us, her intensity razor-sharp.

She told us family stories, and Sylvia was always the star. This sharpened some elements of the past for us, but shadowed others. We heard next to nothing about Pody, Sylvia's younger sister, who also died of ovarian cancer, at 45, several years after Sylvia died. For my mother, family history was a story with only one protagonist. It was a story, not of genes, but of individual tragedy. Sylvia 1.0.

My mother told us different versions. At times, Sylvia's death seemed like a negative version of the Meyer myth. *Sylvia was tired of life*, my mother reflected, sitting at our kitchen table with her ubiquitous pack of Pall Malls and her favorite yellow ashtray. *She was out of money. She hated being in her forties.* My sisters and I looked uneasily at each other. Was she saying Sylvia died on purpose? We didn't know much about cancer, but this seemed wrong.

"She didn't go to the doctor when she should have," my mother concluded, tapping the ash off her cigarette.

Death could be avoided, then. It was just a question of will.

My mother was angry with Sylvia for abandoning her, my father confided once. He was always Sylvia's champion. "You girls would have adored her," he mused.

We mulled this over.

What connected us to the past? What did the stories my mother told have to do with us?

Because it wasn't just Sylvia who died. There was also Pody.

What about my mother? Would she be okay?

In most things, my mother was the apotheosis of organization and advanced planning. She was always first in line at carpool. She packed our sandwiches in wax paper with hospital corners. She wrote beautifully penned letters to hotels in London, planning our first family vacation abroad. She wanted to show us Windsor Castle and Hampton Court, seats of royal power. Before we left, she bought the three of us matching trench coats—blue, brown, khaki. Whatever the occasion, she was ready.

I don't remember ever seeing her asleep. Yet in another sense she was always already writing her own elegy. She hung a ceramic tile up in the kitchen—deriding it, of course—that read, "What Is A Home Without A Mother?" The underlying adage haunted me even as I laughed, trying to mirror her. Motherhood and death were interwoven. Still, sentiment was to be avoided.

What can you do?, she'd ask rhetorically. *Carpe diem*—seize the day. You can't ruin your life worrying.

My father had another view. He'd been reading up on familial cancer, and by the late 1970s, he'd started urging her to have a complete hysterectomy. My sisters and I weren't privy to their "discussions" (my parents, by fiat, never argued), but my mother filtered some of this down to us. "Dad says" or "Dad thinks" would preface her accounts. My mother was firmly opposed to surgery. It was "interfering," she said. What will be, will be. Remember the appointment in Samarra?

Then, in the early 1980s, everything changed. I was 23, studying abroad, when my mother learned her cousin Gail, Pody's only child, woke up one morning, her abdomen filled with fluid. Advanced ovarian cancer. Lightning had struck again, but this time, it was right in my mother's own generation. Gail was only 47, two years younger than my mother was.

She and my father flew into action. My mother had surgery months after Gail's diagnosis. We were kept out of it: I didn't even know about the hysterectomy until it was over. I first heard the news on the payphone in the Middle Common Room of Pembroke College, Oxford, feeding fifty-pence pieces into the heptagon-shaped space to silence the "ping" saying our time was up. She'd gotten the pathology back, and it was "clean," she told me. It was over. She'd avoided the family scourge, just in time.

Gail wasn't so lucky. She died less than two years later, in the same hospital—the same week—that her first grandson was born. Never able to hold him or to learn his name.

In the 1980s, we adopted a new family narrative. Elaine v. Gail. This, too, was a story of volition trumping fate. Medical intervention could save you, but only if you were proactive, if you planned. In the new narrative, you needed top medical care and a schedule. *Get married, have children, and get those things out of you.*

There wasn't space in this new narrative for self-doubt or deviation. My sisters and I all wanted to get married and have children. Sara, my older sister, married young and already had two daughters. She had surgery years before Julie and I, who were still slogging through grad school. We all embraced the new narrative, but like my mother, we kept aspects of the older one. When I met Jacques, just shy of my twenty-ninth birthday, I saw our relationship as "meant to be"—not just a happy accident. The fates would protect us if we just did the right things, whatever those were. In our plans, magical thinking and modern medicine intertwined.

Then, out of the blue, my mother found a lump in her breast, and none of the previous stories made sense anymore.

I was mostly through my doctorate, living in Cambridge, Massachusetts, writing a dissertation on utopias. Star-shaped cities. Spaces of perfection and control. When my mother told me the suspicious lump was malignant, I held the phone away, hearing the slow drag and thump of my heart.

This fit into no story I knew.

My mom puzzled it through. Bit by bit, she worked out a new narrative. She'd gotten this cancer—tiny, "curable"—*instead of* ovarian cancer. An appeasement to some invisible god. *This* was better than *that.* Was she frightened? She said not, after the initial shock. She rallied. She underplayed it. She refused to get a second opinion, to travel to Mayo or to the Farber. She swore us all to secrecy. She chose lumpectomy and radiation—standard of care in the late 1980s—over any more radical plan, and she squeezed her treatments in before her first teaching section of AP History each morning. If any of us acted overconcerned, she lashed back with her saber-sharp wit. She was *fine*. If this was the penalty exacted by some vengeful god, so be it. She'd do her time, move on, and so should we.

So we did. Despite this unsettling new chapter, we clung to our narrative—Family 2.0—of careful planning and medical intervention.

My sisters and I navigated our twenties alternating between terror (we thought of our ovaries as "time bombs") and denial (we were young, nothing could happen to us). My father, drawing on his medical training, registered our family with the Steven Piver Institute in New York and with Creighton University, institutions researching families with apparent hereditary ovarian cancers. My sisters and I were told to have children young, if we wanted children (which we all did, fervently). We should subtract ten years from the age of death of the youngest "affected" family member to find the optimum age to have our ovaries surgically removed. Sylvia created that "benchmark" for us: since she'd died at 43, 33 became the boundary for me between safety and terror. My doctor bumped that number up to 35, but I worried he was being overly generous. The minute I was done having children, I wanted those things out of me.

Time for me became something of an obsession. As I saw it, Jacques and I found each other "just in time." We wanted many of the same things. At the top of the list: *children*. Did I push the timetable? No question. Jacques tried his best to understand, though—numbers man that he is—he was baffled that I turned a guideline (35) into a firewall between safety and death.

I couldn't help it. I had fused my mother's magical thinking with my father's unquestioned belief in medical intervention. Thirty-five became a fixed boundary for me: I couldn't cross that line. As it happens, Jacques and I were extremely lucky, able to conceive our first baby just as planned. Sacha was born when I was 32, the year after we got married. I didn't let myself ask what would happen if it hadn't been that easy. Or if it wasn't easy the next time. Greedy, I wanted a second baby.

I wanted to believe we had it figured out. That planning was all it took.

Then, in the wonderful turbulence following Sacha's birth, we learned my mother's breast cancer had come back. It was in her bones—everywhere. They were recommending Megace first, then aggressive chemo. People could do very well with the treatment, she told us. But I could hear the terror in her voice.

Later, we learned that many of the unusual aspects of her disease are characteristic of BRCA1 breast cancers—yes, her tumor had been initially small, "curable," but it was also estrogen-negative, and when it metastasized, it was fast-spreading, heartbreakingly impervious to any of the available chemotherapy agents. The space of time between her recurrence and her death was terrifyingly short—around eight months. An almost-year that changed the way my sisters and I understood everything. Statistics, actuarial charts, blood work—none of these data constitute "knowing." I didn't "know" what hereditary cancer was until I watched my mother die from it. After that, it was impossible to unknow.

I experienced Sacha's first year of life as if through the figure of speech rhetoricians call "chiasmus," the "crisscross" figure in which the second half of an expression opposes the first. Darkness crossed with light. Disease and death opposing health and life. It was a year of everything beginning and everything ending: Sacha blooming, growing, able to do more and more: grasping, crawling, shaping her first words. And my mother fading, losing weight, color, vitality, being pulled from the world just as Sacha crawled and toddled and eventually—amazingly—pulled herself up on both short legs and stumbled forward into it.

"This stops with me," my mother mumbled, days before she died. I was back home, in Michigan; she was in the guest room, upstairs, her tiny cache of supplies on the bedside table. Pill bottles. Kleenex. Ice chips. It was the last time I saw her.

I was just a few weeks pregnant with Elisabeth, our second daughter. I had just told my mother about it. The new baby would be named for her, we agreed. An *E* name. Edward, maybe, if it was a boy.

If it was a girl—

I can't go over those last hours: it's too painful. But I can hear her hoarse, morphine-thick voice: *This stops with me*. A bitter imperative; a wish.

She had been 59 for only three weeks.

Would it stop with her?

Why couldn't she be right?

Numb with grief, my sisters and I assured each other that her death was a fluke. A lightning bolt. Terrible things happen. But that didn't mean they had to *keep* happening. We had our master plan, we just needed to stick to it. For my part, I would have this second baby, love her with all my heart, and get my ovaries out of my body.

Then I got the letter from Creighton University.

It was early April; Elisabeth was weeks old, napping. The letter reported, in noncommittal, single-spaced prose, that researchers had determined our family had a condition they were calling HBOC—hereditary breast and ovarian cancer. My mother's breast cancer was not unrelated to the cancers that had claimed Sylvia, Pody, and Gail. Not a "this instead of that" but part of the same history, unfolding again, however hard we'd tried to prevent it.

A new narrative, overwriting the others. Family 3.0.

What did this mean? Did we give up our master plan? Was there any point in pursuing surgery if we would only be reducing our risk for one disease and not the other?

For me, the news merely intensified the need to act. Not to "know"— whatever that meant. But to do something. The thing I'd planned.

Weeks after I got the letter, I went to see my surgeon, Dr. Muto. Jacques came with me, and we brought Elisabeth, who was 6 or 7 weeks old, strapped into the plastic carrier we called "the bucket." Sacha was home with our babysitter. I rocked the carrier back and forth with one foot as Dr. Muto looked through my file. Before we set up the operation, he wanted us to know that there was a test I could take that would tell me whether I had a mutation called BRCA1.

It was complicated. The test was relatively new. He compared the BRCA gene to the State of Texas. They've found Dallas, he told me. But there were many other cities still out there. A positive result would be confirming, but a negative wouldn't tell us much.

The test was in its early days. Still—

My mind raced ahead. In the absence of validating information, how would my insurance company react to a negative result? I didn't want any impediments to surgery.

Four out of four women dead, I reminded Dr. Muto, in two generations. From age 43 to age 59. I wasn't comfortable waiting anymore.

I was terrified of leaving Sacha and Libby motherless.

Some nights I woke—drenched, heart thumping, thinking I heard
one of them crying—only to silence, the blank wall of our bedroom facing me. I felt awash with the sense they needed me and I wasn't there.
Only later would I understand that the emptiness I feared, the palpable
sense of being without a mother, came from within me. I was the one
without a mother. The line between my mother and me had been perma-
nently blurred; I would forever carry parts of her within me, aware of her
as I raised my voice or sang a fragment of song. Aware that her foreshort-
ened life could limit my own.

Dr. Muto was patient with me. Actually, it *could* matter, he pointed
out. If I tested negative—

He and Jacques looked at each other, past me. Eyes meeting. Jacques
wanted a third child. We were harried, working parents, our lives and
schedules crowded even before the specter of grief edged its way in. A
negative test could give us time.

Except, for me, it couldn't. The more we talked, the clearer it became
that a negative test result would (at that stage) tell us next to nothing.
Neither of my sisters had been tested yet, nor wanted to be. I had two
cousins in Chicago I barely knew—I didn't even have phone numbers for
them. Until you get a positive result, Dr. Muto concurred, mulling it over,
you won't know for certain what a negative means.

That settled it. Six months later I had my ovaries removed. Julie, my
younger sister, was just behind me. There wasn't time to worry about
the test.

Three years later, when we were weighing the pros and cons of prophylactic mastectomies, we came back to the question of testing, this
time with greater intensity. A second mutation had been discovered in the
intervening years, and in some ways, that intensified our reluctance. If
there was a BRCA2, there could easily be a BRCA3 or 4, we reasoned.
Right now, we had our families and our insurance companies on board
with this second surgery. We had momentum. A negative test result could
impede that.

Or so we said.

One by one, we had preventive breast surgery. I've put some parts of this experience out of my mind, but memories come back sometimes, in a shadowy, ill-lit way. Lying on the gurney in Mass Gen, next to a medical supplies cupboard, my teeth chattering, vision blurred because I didn't have my glasses on. Someone laid a clipboard on me. A nurse came over, squeezed my hand, eyes flicking compassionately over me. "Poor thing," she murmured, "you're young for this, but don't worry, they're good at getting these things, and I bet they've caught it early."

I stared up at her, tongue-tied, unwilling to take the tenderness she was offering and even less willing to go into the OR with a lie hanging over me. "I don't have cancer," I mumbled, and she pulled her hand back with the recoil of a gun. Just like that. Pity turned to . . . what? Confusion? Something worse?

*What kind of person would do this to herself?*

When I talk with my students about medical practices in the seventeenth century, they are horrified by what constituted "healing" in those days. Blood-letting. Peculiar potions. Applying leeches (which in some cases has come back in favor today). It's easy to see another age or culture as barbaric, to believe wholeheartedly in progress—that each step we take is a step forward. It's harder to admit that we fumble, each of us, hoping we have it right. Sometimes moving forward, sometimes not. When I woke from surgery to a pain too intense to articulate, vomiting bile, I clung to a single idea: I was done now. I had done everything possible, and I could put this behind me and live whatever life I had left. Maybe not without fear, but without recrimination.

Was it partly because my sisters and I chose surgery before (or instead of) testing that our conversations about the test itself became increasingly complex?

By 1997 there were tests available for BRCA2 as well as BRCA1, but nobody in my family had yet been tested. My sisters and I talked it over in intricate detail. We worried, all of us, about the implications—that it might give us false reassurance, keep our eyes off the goal, which we saw as surgical intervention.

We worried it might divide us from each other.

We worried, before the Genetic Information Nondiscrimination Act (GINA) was put in place, about discrimination.

For my part, I worried about the costs of knowing. It was one thing to go through surgery, to reduce my risk. It would be another, I felt, to  know for sure. Why? Because there are other risks I can do nothing about? Maybe. Increased risk for colon and pancreatic cancers, for melanoma. I wanted, if possible, to minimize risk and worry at the same time.

Later, Sara got tested. She was negative. I reached out to my cousins in Chicago—Gail's children. They had been tested. Paul was positive; his sister, not. So we finally had the data we needed—our broken gene. Now, what I learned from testing would be "informative," my doctors told me.

As simply as I can put it: I didn't want to know. When Sacha and Elisabeth asked if I had the mutation, I was able to tell them—honestly— that I wasn't sure. I wanted both, the honesty and the uncertainty. I still want both.

Looking at Francis Collins's algebraic formulation, I wonder where to put the symbol for ambivalence. For guilt. For the longing to stop time, to hover, for a bit longer, in the space of indeterminacy.

This year, I am the same age my mother was when she was first diagnosed with breast cancer.

At 54, it's hard to see what lies ahead on the road, or which path to take. How to direct my daughters to live their lives judiciously, wisely, yet with freedom and joy. "Getting tested," the phrase that has become shorthand in our vocabulary for knowing and accepting the truth, looms ahead for me, much closer than it was a few years back. For my daughters now, as well as for me. But I also know that's only part of what follows. I'm fairly certain there will be new narratives to replace the one we live with now. Family 4.0 and 5.0, and more. My daughters will help to write these new stories, even as they get written around them.

But even in these new, updated versions, older myths remain. What you learn as you get older is that time runs in at least two directions at once. Forward, to places we can't see, and backward, to spaces of vacuity and shadow. Jacques and I, lifting Elisabeth in our arms. Calling her Libby, to free her from that E even as we honor it. Before her, Sacha, eyes fixed soberly on me as my mother lifted her out of her crib, showing me (the way she always did) how. Back, way back. My mother, leading my sisters and me through the maze at Hampton Court, glued to her guide-book, searching for the opening. Sylvia and Pody, laughing on some un-named beach, faces flushed with girlhood. That window in Lincoln Park, where Meyer, stroke-bound, flies backward, up and over the sill, back to his bed, to health, humor, and vigor, back to the boat, and back to a tiny, unnamed village, where he unrolls his sock, presses coins into his moth-er's outstretched hand, his goodbyes muffled in the warmth of her as she pulls him close to her, as if she'll never let go.

# FIVE

# The Unnumbered

Alice Wexler

....................................................................................

In 1952, the Jewish Bulgarian-born writer Elias Canetti wrote a satiric play, *Die Befristeten* (*The Numbered*), set in a dystopic future when everyone knew the number of years they would live. In the opening scene, two men converse about "the old days" when no one knew when their "moment" (of death) would come. The men can hardly imagine such a barbaric time. "I don't understand," one says to the other. "You're trying to tell me that no one, not one single person, had any idea of the moment at which he would die?" In their view, "no one could have stood it, such uncertainty, such fear. I shouldn't have had a moment's peace. I should have been able to think of nothing else. How did those people live? One couldn't have taken a step out of one's house—How could people make plans, how could they embark on anything? I think that's terrible." The men are convinced of their own superiority. As one puts it, "Any wretched cobbler now is a better philosopher because he knows what will happen to him. He knows when he will die. He can apportion his lifetime. He can plan without fear. He is sure of his allotted years and he stands as firmly on them as on his two feet."[1]

I was astonished when I first read *The Numbered* in 2009.[2] Although it was written shortly after World War II, it seemed so contemporary! For me, the play also had a peculiarly personal resonance. The notion that everyone knew the number of years they had to live and that their birth and death dates were recorded in sealed lockets worn around their necks recalled something completely unknown in Canetti's time and seemingly far from his concerns—namely, the early discourse of predictive genetic testing for Huntington's disease, a fatal hereditary neurological and psychiatric disorder associated with an excess number of CAG repeats in the gene for a protein called huntingtin. The idea that knowing one's genetic future—and knowing how many CAG repeats one possessed—could reduce anxiety and uncertainty and help in planning one's life: these notions promoted in the early literature on HD testing seemed straight out of the opening scene in *The Numbered*, albeit without Canetti's irony and satirical intent.[3] Long before I discovered Canetti's play I had published a memoir, *Mapping Fate*, about the research leading to the discovery of the fateful CAG numbers. I had written about my family's experience, both as participants in this research and as people living with Huntington's, and about my own struggle, and that of my sister, to decide whether to get tested ourselves and learn our numbers.[4] The idea of "the numbered" was overwhelmingly real.

Formerly called Huntington's chorea, Huntington's disease today is defined as a degenerative neurological, psychiatric, and motor disorder emerging typically in the thirties or forties and causing irreversible decline over ten to fifteen years. Symptoms include jerky involuntary movements called chorea, personality changes, and cognitive losses leading in many cases to dementia. Medication can allay depression and ease chorea, but as yet, there is no intervention to delay or prevent the onset or progression of the disease. Since Huntington's is transmitted as an autosomal dominant disorder, each child of an affected parent has a 50 percent risk of inheriting the malady and of passing it on to his or her children. Moreover, unlike recessive disorders that require both parents to be carriers, a single copy of the abnormal version of the HD gene can confer the illness. As of 2012, the Huntington's genetic test is one of very few that can indicate with almost 100 percent certainty that you will *not* get HD or that you *will* get it if you live a normal life span. The numbers are critical, for one additional CAG repeat—40 instead of 39—can spell the difference between "normal" life and the disease.[5]

For this essay, I decided to look back on *Mapping Fate* through the lens of *The Numbered*. I wanted to reconsider the meanings of knowing or not knowing certain numbers—a 50 percent risk, 100 percent certainty, a range of reduced penetrance, a precise number of CAG repeats. How were certain numbers first hidden and then highlighted in several generations of our family history? Is genetic information such as having 42 CAG repeats more telling for one's sense of identity than other kinds of medical information, as the psychologist Kimberly Quaid and many others have suggested?[6]

## Finding Out about Huntington's: The Past

Though I had begun *Mapping Fate* out of a desire to document the research leading to identification of the HD mutation, I found that I also wanted to write about the emotional impact of Huntington's in our family. Huntington's, in fact, gave me a sense of legitimacy in writing a highly personal memoir. I felt that I was writing about a critical moment in the history of genetics as well as exploring my own life. Huntington's also gave me a kind of algorithm for the memoir: I would write only about those aspects of our family life that related to Huntington's. I identified with Chinese American writer Maxine Hong Kingston in *The Woman Warrior* (1975) when she asked, "how do you separate what is peculiar to childhood, to poverty, insanities, one family, your mother who marked your growing with stories, from what is Chinese?"[7] What in our lives, I wanted to know, was related to being Jewish, growing up in the 1950s, my mother's loneliness, my father's infidelity, his profession of psychoanalysis, and what was related to Huntington's disease?

Finding out about Huntington's had marked a sharp divide in the lives of my sister and me. In 1968, a year of dramatic political and social events around the world, our mother was diagnosed with HD, at the age of 53, and we suddenly learned we each had a 50 percent risk of developing it, too. After she learned of her own diagnosis, my mother told us she had known about Huntington's ever since her father died (in 1929) when she was 15: she had looked it up at the library. But she claimed that she had thought only men got this disease. Indeed, in her family, until her diagnosis, it had affected only men: her father, and then all three of her brothers. She claimed—although I did not believe her—that since she was convinced women did not get it, she had thought she was not in danger of developing it or of passing it on; no need to tell my father about it when they married or to worry her daughters, since it would not affect us. Although she and my father learned in 1950 that women could get Huntington's, they still did not tell my sister and me, not until Mom was diagnosed herself, eighteen years later. All those years they had kept

Huntington's hidden from my sister and me. I was shocked and furious that they had not told us, even as I tried to sympathize with my mother as she began to confront the reality of her illness.

By the early 1980s, the more hopeful atmosphere surrounding Huntington's, created by the emergence of HD advocacy (spearheaded by my father and sister and by the widow of Woody Guthrie), made it easier for me to consider the painful past. My mother had died in 1978, ending the cycle so far in our family. I began to trace the transmission of both knowing and not knowing in our family: the silences, denials, truths, and fictions passed down along with the disease. Huntington's made sense of certain mysteries that had been part of our family memory. For instance, it explained my mother's reluctance to get married and my maternal grandmother's peculiar behavior on the eve of my parents' wedding, when she urged my father to telephone a certain physician, who hinted at confidential information he could not divulge. It helped explain my mother's transformation from an intellectually lively young woman, one who had earned a master's degree in zoology in 1934 and loved teaching high school biology, into a passive and depressed middle-aged housewife. It helped explain why we did not see our New York relatives more often and why my Uncle Seymour, who did visit us once or twice in Los Angeles, kept winking and blinking and jerking as he showed us his magic tricks. The process of writing eased my anger at my mother's passivity and lessened the guilt I always felt on her behalf. I came to appreciate how the early twentieth-century eugenics movement could have contributed to her sense of shame and guilt. She had grown up at the height of organized eugenics, in the 1920s and 1930s, and was surely exposed to ideas about the "fit" and "unfit" in the media. She would have studied biology in books such as eugenics leader Charles B. Davenport's widely used textbook *Heredity in Relation to Eugenics* (1911), which implied that people like her were "defective" and should never have children.[8] How could she not have internalized such stigmatizing ideas? Writing about the ways in which Huntington's had shaped our family life helped me let go of the rage I had felt toward both my parents and made me more sympathetic to my mother's difficult life.

Years after *Mapping Fate* was published, I also found that some of what I blamed on Huntington's in the memoir did not quite fit the historical time frame of events. For example, I had traveled to Argentina in the summer of 1967 for preliminary research on a PhD history dissertation focusing on the tango and popular culture in early twentieth-century Buenos Aires. In the memoir I wrote that the shock of learning about Huntington's in the summer of 1968 had shattered my plans to return to Argentina and led me to change my entire research focus (and field of specialization) in order to remain in the United States, close to home. But recently, when I reexamined letters and diaries from the 1960s, it became clear that I had changed my topic months *before* my mother's diagnosis, in the fall of 1967, soon after I came back from Buenos Aires. Huntington's evidently had nothing to do with it.

The reasons for this misremembering are unclear to me still. Perhaps I did not want to recall that I had felt afraid to return to Argentina now that the generals were taking control. I preferred not to remember that I had felt lonely living in Buenos Aires on my own. I did not wish to acknowledge that my major professor, whom I idolized, had not been pleased with my progress when I returned or that I, too, felt dissatisfied with the proposal I had developed. But my misremembering may also have related to a harsh letter of (non)-recommendation from this professor that I had discovered in my tenure file sometime in the late 1970s, though it was written in 1971 when I was starting to look for my first academic job. That letter had been so painful that for years I could not bear to think about it. In it, my professor wrote that he doubted I would become a productive scholar-writer, since I had changed my dissertation topic several times and clearly lacked the necessary "staying power" and discipline required in this field. My record, unfortunately, was "somewhat spotty."[9] (I had not told him about HD.) In transposing the date of my shift in focus, I could blame my inadequacy (for that's how I perceived the import of his letter) on the discovery of HD in the family. (Somehow, blaming genetics rather than gender felt easier, even in the 1990s!) As the historian Alessandro Portelli writes, "memory goes to work to heal" the wound, in this case the wound of my professor's letter.[10] Huntington's seemed a legitimate reason to have abandoned the tango dissertation and Latin American history and to have floundered for a while among different topics. The narrative of the memoir was a rebuke to my professor for his job-busting letter, even if it did not fit the facts.

## Finding Out about Huntington's: The Future

Finding out in 1968 that our grandfather, uncles, and now our mother all suffered from Huntington's disease meant that my sister and I had acquired a new identity. We now knew we had a 50 percent risk of inheriting Huntington's but also an equal chance of escaping it entirely: 50-50. The flip of a coin. We had become persons "at risk for HD." Researchers have found that people have a hard time keeping a 50 percent risk in mind. Rather, they tend to embrace either the 100 percent likelihood that they *will* get HD or the 100 percent likelihood that they will not.[11] I focused on the latter while still anxious about the former. Mostly I tried to put the illness out of my mind, except when I visited my mother. For the next decade and a half I avoided all things Huntington, even as my father and sister were becoming leaders in an emerging HD advocacy movement.

Once researchers (including my sister) located a genetic marker for HD in 1983 and the genetic mutation itself was on the horizon, everything changed.[12] I found myself swept up in the euphoria of the moment. We even believed that a cure might be just over the horizon. It was at that point that I began to think about documenting this research. In 1986, a few academic medical centers in North America and Europe began offering the predictive genetic (linkage) test in a research setting. In 1993, my sister and her team of researchers identified the abnormal version (or allele) of the HD gene itself, that is, the allele with the expanded stretch of CAG repeats. At that point, predictive testing became technically simpler and more accurate and began to be offered in many more centers. In the diary I had kept for many years, I recorded our family discussions about testing and my own ambivalent thoughts and fears. My sister and I had lived for fifteen years knowing our 50 percent risk. Did we want 100 percent certainty that we would or would not develop HD? What if we learned we were among those rare individuals who had CAGs in the "high normal" or "reduced penetrance" range and were left with even more uncertainty than before?

In the 1980s and early 1990s I was in my forties. I was living with a long-term partner. We had no children. I was not sure I wanted any, although

I began trying (unsuccessfully) to conceive. I felt my entire life suspended at the edge of a genetic precipice. Any morning, a policeman might stop me in the street, scolding me for being drunk, even if, like my mother in 1968, all I had had to drink was a coffee. If I took the test, I would finally know one way or the other. Would my anxieties diminish or would they increase? If I had inherited the expanded version of the gene, presumably I would die with HD and at a younger age than if I had not inherited it. If I got pregnant, I could pass it on. Would the new information be a comfort or a catastrophe?

I reminded myself that half the people tested would presumably learn that they were *not* going to get HD and would be relieved of an enormous burden. I might be one of them. The other half, at least, would be able to plan their lives accordingly. Or so the early literature on predictive testing seemed to indicate. Writing the memoir became an occasion for exploring these issues. As Joan Didion famously put it, "I write entirely to find out what I'm thinking."[13] Through the process of writing, I found myself slowly embracing the identity of a woman at risk and the 50-50 uncertainty it entailed. To my surprise, I found that I enjoyed the "third position," as feminists and other theorists sometimes classified identities outside the common binaries of Western culture, such as male/female, black/white, masculine/feminine, heterosexual/homosexual, or in this case, gene-positive/gene-negative. In the academic culture of the 1980s and 1990s, when binary thinking generally was under attack in many quarters, I flaunted my ambiguous genetic identity. I liked being on the border, so to speak.

I decided not to get tested. At the time, many people called this choice a form of denial. Indeed, the notion that taking the test was facing "the truth" while choosing not to get tested was a way of avoiding "the truth" circulated in the popular media as well as among some HD family members and even some clinicians. As an article in the Minneapolis *Star-Tribune* in 1995 noted, "Genetics research has led to the development of tests for serious, sometimes incurable, illnesses, creating the dilemma of whether to take the emotional and financial risks of learning the truth."[14] Medical students to whom I spoke in a medical history class at the University of Southern California were scathing about my decision not to get tested, as I had described it in *Mapping Fate*. As the class professor,

Philippa Levine, later wrote, "My students largely condemned Wexler for choosing ignorance over knowledge, as they framed the question. Many of them thought it was Wexler's duty as an intelligent and informed woman to further the cause of science and medicine and to set an example to others."[15] Some of them seemed to believe that new technologies should be used simply because they were available.

I argued that two different kinds of knowledge were at play—knowledge of the 50 percent risk and knowledge of a 100 percent likelihood that one did or did not carry the genetic mutation—and that both were equally legitimate forms of knowing. I tried to persuade the students that the number of one's CAG repeats might be accurate information but was not necessarily useful "knowledge," since it did not predict when symptoms might begin or how severe they might be over a long period of time or how one might live *before* or even *after* the symptoms emerged. So long as there was no *medical* benefit to learning one's genetic status and a reasonable likelihood of stigmatization and discrimination, the choice for or against acquiring this information was purely existential. I reminded them that stigma and secrecy had a history, and I spoke of the historical eugenics movement with its exclusions, sterilizations, and even exterminations (under Nazism). But the students remained unmoved.

On the other hand, some people in the HD community appreciated my efforts to legitimize the choice *not* to get tested. Drawing on diaries I had kept at the time, I had written extensively about our struggles in order to show that even those with considerable resources and social support, such as my sister and I, found this decision immensely difficult. I hoped that revealing the difficulties we experienced might offer comfort to others going through similar struggles, often without the support that my sister and I had.

Today, even the popular media accord more recognition to the complexities and perils associated with predictive genetic tests.[16] It is telling that, in 2009, when Harvard psychologist Steven Pinker, an enthusiastic proponent of personal genomics, got his genome scanned as part of the Personal Genome Project at Harvard, he chose not to learn whether he carried a mutation that heightened his risk for Alzheimer's disease. "I figured that my current burden of existential dread is just about right," he wrote in the *New York Times Magazine*, "so I followed [James] Watson's lead and asked for a line-item veto of my APOE gene information when the PGP sequencer gets to it."[17]

I have still not learned the number of my CAG repeats, although any day I may decide to do so. As of 2000, worldwide, a small minority of adults at 50 percent risk for Huntington's disease (a majority of them women) had chosen to do so.[18] But whether or not we choose this path, the numbers do undeniably exist. Our lockets are not empty. However, unlike those in Canetti's play, our numbers do not predict our "moment." Indeed, one of the major findings from the past decade of HD research is that individuals with precisely the same number of CAG repeats in their huntingtin gene—for instance, 42—may develop symptoms of Huntington's disease at dramatically different ages, due to environmental and epigenetic factors (mechanisms that influence gene expression). The identical genetic mutation can have radically different outcomes and effects. Even so genetically determined a disorder as Huntington's is not entirely determined by a gene.

## Epilogue

Toward the end of *The Numbered*, the character named Fifty decides to open the sealed locket that he—like everyone else—wears around his neck, containing the dates of his birth and death and therefore of his "moment." But instead, he finds that his locket is empty. He opens the lockets of several others and they, too, are empty. The Keeper of the Lockets, the only person authorized to open anyone's locket, and only after the individual's death, turns out to be a liar and a fake. Society's entire system of belief in "the moment" is revealed as fraudulent. But even though many people confide that they were unhappy with the knowledge of their "moment," they now find themselves tormented by the new experience of uncertainty in the face of death. Even Fifty, "the Deliverer," has his doubts.

And yet, today, some of those people who know they have the extra CAG repeats are using that information in exciting and inventive ways. They are writing profoundly about the meaning of living with their "number" and with the certainty of future illness. They are making films and writing novels and making dances and video games. They are collaborating with visual artists, writers, animators, actors, theater directors, philosophers, and choreographers to create alternative understandings of HD. They are creating new knowledge "from the perspective of the Huntingtonian."[19] These awesome young people around the world have put their sense of urgency, their imagination, their fantasies, even their anxieties, into powerful forms of communication. Some have become scientists themselves, working in laboratories to develop the treatments that may one day change our lives.[20] They know their numbers, but their numbers do not define them. Their numbers may be given, but their "moments"—not of death but of life—remain exhilaratingly unknown.

## PART TWO

# Intervening

*Living with Genetic Difference*

# SIX

# Of Helices, HIPAA, Hairballs . . . and Humans

Misha Angrist

. . . . . . . . . . . . . . . . . . . . . . . . . . . . . . . . . . . . . . . . . . . . . . . . . . . . . . . . . . . . . . . . . . . . . . . . . . . . . . . . . . . . . . .

In early 2006, *Scientific American* ran a cover story entitled "Genomes for All," written by George Church, a prominent geneticist at Harvard.[1] Church envisioned a day in the near future when complete genome sequences would be affordable for most people. So, he wondered, how exactly were we going to go about this?

For me, the question itself was a revelation. Church was saying what I had felt for years but could never articulate: the time for genetic exceptionalism—the idea that genetic information is special and should be treated differently than, say, cholesterol levels or cancer diagnoses—was over, and the time to include large numbers of ordinary volunteers in the act of genome sequencing had arrived.

To that end, Church launched the Personal Genome Project, a visionary experiment in which informed citizens could have their genomes sequenced for free if they consented to make the results—and their medical information—available for public use. I badgered Church until he let me into the pilot group of ten participants, which included, among others, Church himself, Steven Pinker, and Esther Dyson.[2]

Subsequently, I had my complete genome sequenced—most of the six billion DNA letters in my cells—but for all my excitement about the project, my results turned out to be something of a nonevent, as they have for most otherwise healthy people in the PGP and elsewhere. As a card-carrying genetics geek, I was captivated by both the enormity of my sequence and the minutiae. I browsed my favorite genes and was awestruck by how much of the variation in my genome had never been seen before. It turns out we all carry thousands of variants that are rarely if ever seen beyond our own blood relatives, and sometimes not even in them.[3]

But what if I carried a known disease-causing variant? That information would now be a mouse click away for anyone who cared. Participating in the PGP meant putting my money where my mouth was. Was I really prepared to share probabilistic information about my and my family's genetic predispositions?

## That Heeds No Call to Die

If I were more self-assured, I'd spare my students the lengthy, portentous syllabus I distribute for the freshman seminar I teach at Duke, "Secrets of Life: DNA, Property Rights, and Human Identity." Instead I'd let Thomas Hardy's 1917 poem "Heredity" do the work.

> I am the family face;
> Flesh perishes, I live on,
> Projecting trait and trace
> Through time to times anon,
> And leaping from place to place
> Over oblivion
>
> The years-heired feature that can
> In curve and voice and eye
> Despise the human span
> Of durance—that is I;
> The eternal thing in man,
> That heeds no call to die.

## Determined to Be Exceptional

Genetic information has long been regarded as both exceptional and deterministic, as special, powerful, predictive, scary; information that must be kept secret at all costs.[4]

No one openly admits to genetic determinism. No one congratulates Justice Oliver Wendell Holmes for his 1927 declaration in support of forced sterilization: "Three generations of imbeciles is enough!" No one I know sings the praises of *The Bell Curve.*[5] Most of my colleagues would regard being called a genetic exceptionalist or determinist on a par with being challenged to pistols at dawn.

But the truth, I'm sorry to say, is that most of us actually are genetic exceptionalists/determinists. Not because we're intent on culling genetic "defectives," but because we've been exalting heredity for so long we don't know anything else, and because we believe it's in our own interests to keep thinking that way. It's not just genetic information that's exceptional; exceptionalism has seized upon all human biomedical research.

## Toiling at the DMV

I'm deeply unhappy with the anonymous way human research participants (I refuse to call them "subjects") are treated by the biomedical research enterprise. Trying to foment a little change from within, I serve on the Duke Health System's Institutional Review Board, whose thankless charge is to review and critique research protocols submitted by campus researchers ("investigators") to ensure that their research programs will be or have been carried out in an ethical and just fashion. We vote to "approve," "approve with modifications," "reject," or "defer" each protocol.

Recently I reviewed a biobanking protocol. The goal was to consolidate forty thousand existing blood and tissue samples and accompanying data with prospective collections of similar samples. Creating biobanks is sensible: human tissue (and its DNA) is essential to understanding disease and developing treatments. In this protocol, patients undergoing a surgery or diagnostic biopsy would be asked to give a blood sample and accompanying health information. Much—perhaps all—of this material would then be farmed out to investigators doing genetic research.

But what were participants being asked exactly? One of my colleagues on the IRB read the biobank consent form and, with a mixture of disgust and grudging admiration, said, "It looks like it's been gone over with a fine-toothed comb by Duke's best lawyers." I'm sure it had. Paraphrased, the consent form said the following:

"Subjects" will get no compensation for participating.

The biobank will try to profit from the samples, offering them for sale.

"Subjects" will consent to be in the biobank in perpetuity.

"Subjects" will get no access to any of the genetic or other results generated by any of the researchers using their samples.

If "subjects" were to get injured in any of the studies, Duke would treat them, but the bill will go to them or their insurance company.

Last but not least, the consent form assured them that "subjects would be giving up none of their rights."

These provisions are not atypical, but that doesn't mean I like them.[6] I think when we ask participants for broad consent to use their tissue and medical data in perpetuity for who knows what, we owe them something. In this spirit, I wrote to the principal investigator and said so. I asked him if he would consider adding the following sentence to his consent form: "If you wish to receive an annual notification of which research groups are studying your samples and a brief description of what they are studying, please provide an email address at the end of this consent form."

He answered right away, very graciously. We went back and forth a few times. I wanted him to know that I felt his pain but that I was convinced that some good would come out of this: "I recognize that making these communications personal demands a little more from you, but I would argue that it's exactly that piece—the personal acknowledgement—that would add value to the participants and ultimately redound to you and the repository."

I waited for his next response.

I am still waiting.

Meanwhile, I took this idea back to the IRB: I told them that I'd like this researcher to send biobank participants an email once a year thanking them and letting them know—in very general terms—what was going on with their samples. "Just a few sentences," I said. The oxygen seemed to leave the room; it was as if I'd asked to add a "kicking of puppies" provision. A kerfuffle ensued.

"This will set a precedent!" said the vice-chair, a retired Southern gentleman physician who I really like and to whom the younger chair often deferred. "Exactly!" I said. He was agreeing with me! Wasn't he? "We can't do this!" he continued. "You're going to bring all research on humans at Duke to a halt!" he said. "If they don't like the protocol, they can choose not to be in the study!" said another member. And my favorite objection: "This poses a HIPAA problem! If you send emails to them, these people are going to be *identifiable!*"

HIPAA—the Health Insurance Portability and Accountability Act— includes a privacy provision intended to limit the disclosure of an individual's "protected health information." The statute is much reviled and, at least until recently, poorly enforced.[7] By advocating that researchers disclose a modest bit of health information *about* individuals *to* those same individuals (and no one else), I was, in the IRB's eyes, advocating violation of those individuals' privacy.

I could see smoke coming out of my colleagues' ears—IRB meetings run three or four hours already and here I'd tossed a smoke bomb into the room, making a long Thursday afternoon even longer. But a few light bulbs were going on, too. We'd already agreed that the consent was too broad. Someone conceded, "Well, you know, informing participants is consistent with the ethical principle of autonomy." Another—a community representative—admitted, "This consent form is complete and accurate . . . but *I* wouldn't sign it!"

When the fracas died down, we took a vote on whether to approve the protocol without any of my proposed bells and whistles. The tally: ten in favor of approval, six against, and two abstentions. I was shocked. And delighted: the motion lacked the 75 percent supermajority it needed to pass! As the IRB Board Specialist concluded, "This has never happened before." We looked at one another.

Then, just as quickly, the dream was over. The vice-chair demanded that the naysayers justify their position and offer the researcher some realistic guidance about his protocol. Another vote was taken: sixteen to two in favor of the protocol. One other quixotic soul and I were the last holdouts.

Under the present rules in the United States, it's easier to keep genetic information, and in fact *all* research information pertaining to human beings, a secret. The law says that only special labs can disclose the data, and even if it's about you and you paid for it with your tax dollars, that doesn't mean that you have a right to *see* it.[8] That's not always the case: a quasi-enlightened protocol I reviewed a few years ago included provisions for returning everything to participants. Everything, that is, *but* genetic information, which, naturally, was deemed too scary to reveal to mere mortals.

Admittedly, the incentives for investigators to keep silent are seductive. As a researcher, if I "de-identify" you and therefore your protected health information cannot be associated with you specifically, then I'm off the hook. I can get an exemption from the IRB and do my experiments on your samples with impunity.[9] Even better, if I don't know who you are then I don't have to talk to you. Ever. Never mind that it's not clear that DNA information can ever really be anonymized, since it is a very long digital code that is different in everyone on earth except identical twins.[10] The point is, I can maintain that I am doing this for your own good: if you're anonymous, then no one can discriminate against you or steal your identity or clone you. The presumption is that if I have your DNA, then I can do all of those things easily.

Except actually, I can't. Genetic discrimination is rare in the United States.[11] And while existing antidiscrimination legislation has some real problems, genetic discrimination is indeed illegal.[12] I imagine it's something people worry about because they've seen the film *Gattaca*, but to be honest, there are much easier ways to discriminate against people without the awesome power of molecular biology. Like, for example, just by *looking* at them. A few years ago, a human resources vice-president at Wal-Mart said in a leaked memo that the company should stop hiring fat people because they're more expensive to insure.[13] Abhorrent, to be sure, but probably true and certainly a lot less cumbersome than sequencing employees' obesity genes.

Identity theft, too, is easier done via credit card pilfer or Facebook hack than by surreptitiously lifting someone's DNA and trying to use it in nefarious ways. And cloning a human takes more than a few test tubes and a maniacal laugh. But these are the fears that smolder in the popular imagination, and nearly every lawyer worth his fiduciary salt is happy to fan the flames.

Greasing the skids for research is one incentive to exalt genetic information and keep people away from it. A more obvious one is money. The
genome fever that overtook the field during the heyday of the Human
Genome Project was stunning. Unlike, say, the space program with its
built-in mass appeal to adventure, the genome was an abstraction that
had to be sold: to the many biologists who saw genomics as a threat to
their own research,[14] to an American public who didn't understand it,[15]
and especially to Congress, which had the power to fund it lavishly or, if
it chose, to smother it in the crib.[16]

The genomics community marketed the HGP with unbridled zeal.
In 1989, James Watson, DNA co-discoverer and first director of the
government-sponsored HGP, said, "We used to think our fate was in our
stars. Now we know, in large measure, our fate is in our genes."[17] One of
the HGP's scientific architects called the HGP "a medical revolution that
holds the prospect that our children's children will never die of cancer."[18]
Francis Collins, Watson's successor as leader of the government-sponsored
HGP (and now NIH director), said that by decoding the human genome
we would be reading a book written in "the language of God."[19]

This kind of exceptionalism and determinism was good for business.
Not the old fear-mongering business of eugenics, but the shiny, new, and
enlightened business springing from the medical-industrial complex (a
teat at which, it must be said, I continue to feed). The 1990s saw biotech
investment (and rabid speculation) skyrocket,[20] and what geneticist Jim
Evans calls "the genomics bubble" inflate to dot-comical proportions.[21]

But science—unlike, say, investment banking—must ultimately stand
or fall on reality as defined by peer review and independent replication.
And genomic reality has turned out to be complicated and, so far at least,
disappointing. We know God's alphabet just fine (DNA is only four letters
strung together in endless combinations), but He or She seems to speak
largely in riddles.

Consider height. From a geneticist's point of view, human height is a
great trait to study: it is highly heritable (80% to 90% of the variation in
human stature can be attributed to genes), it's not terribly controversial,
and it's easy to measure. For those reasons, we have *tons* of data on the

subject: to date, the genetics of height has been studied in well over 180,000 people. So what do we find? It turns out that height is the poster child for the absurd messiness of human heredity: it takes 180 genetic markers to explain just 12 percent of the heritability of human height. To explain half of the heritability requires more than 300,000 markers.[22]

This is fascinating, to be sure, but also sobering: to make predictions about human stature based purely on DNA quickly becomes a hopeless prospect. It is the equivalent of genetic pointillism: each genetic marker is a dot of green on a blade of grass in a Georges Seurat painting—collectively important, but individually meaningless.

By 2009, the genomics bubble had burst. It had become clear that within the foreseeable future, there would not be reliable, deterministic genetic predictors for height, Crohn's disease, early heart attack, multiple sclerosis, bipolar disorder, lupus, or body mass index (among many other common traits). If I want to know how tall my daughter will be, I am better off collecting height data from everyone else in the family or extrapolating from her own height at a younger age. If I want to make an educated guess as to whether I will develop type 2 diabetes, my best bet is not to send my DNA to a lab but to find a tape measure and put it around my waist.

There are exceptions to this, of course. Rarer, single-gene diseases like cystic fibrosis and sickle cell anemia tend to follow Mendel's simple rules of inheritance in more predictable ways. Each of us carries at least a few broken genes that could put us at risk for such conditions. Depending on the genetics of our mate, we could be putting our children at risk as well.

For most traits, however, treating DNA as an infallible crystal ball is unlikely to get us very far. Yes, there are cases where genomics is useful in everyday human health: understanding likely responses to certain medications based on genes, for example, has a real chance to change medical practice.[23] Elsewhere, some larger genetic rearrangements seem to explain a certain fraction of psychiatric disorders.[24] And there is a common genetic variant that has a fair amount to say about the development of Alzheimer's disease late in life, knowledge of which most people are emotionally equipped to handle.[25]

But even rare and formidable broken genes don't operate in a vacuum. If there were any doubts, recent data about the functional importance of what we used to call "junk DNA" (a tribute to our own ignorance) are proof positive that the biology of human life usually turns out to be more complicated than we could have imagined.[26]

So where does that leave us?

My hope is that it will leave us more sanguine about the human genome and what it can and cannot tell us. There are attendant concerns, of course: Disclosure of non-paternity accidentally discovered during genetic testing is always going to be uncomfortable, if not a bad idea. The exponential growth in forensic DNA databases is indeed worrisome, as is the continued willingness to allow surreptitious collection of DNA from objects like an abandoned Starbucks cup.[27] In these contexts, DNA is, like fingerprints, a tool, albeit a precise and potentially potent one. If we abuse it, it will inevitably come back to bite us.

But as human genomes become cheaper to sequence, and sequence data more ubiquitous, I'm less convinced than ever that averting our eyes from what's inside our own cells is possible, let alone helpful in influencing how it gets used.

In the meantime, I'd like my well-meaning but paternalistic colleagues on the IRB to take a deep cleansing breath and stop treating genetic and other health information like unexploded incendiary devices. This goes for physicians, too. I'd like researchers to remain in contact with the people they study and to manage their expectations about results, rather than pretending they don't exist. When possible, I'd like those researchers to share what they learn with the people who are generous enough to donate their time, taxes, and tissues. And I'd like scientists to think twice before describing their research with words like "game change," "cure," "transformative," and even "personalized medicine." Ultimately, history and empirical evidence will make those judgments, not us.

## Envelope Please

When I had my own genome sequenced, the risk about which I fretted the most was for hereditary breast cancer. My mother was diagnosed with breast cancer at age 42 (she is alive and well and wondering why you never call) and is of Ashkenazi extraction. This meant there was a decent chance that she carried one of the more common "Jewish" mutations in the BRCA1 or BRCA2 genes that predispose to hereditary breast and ovarian cancer. While it was unlikely (but not impossible) that I, a man, would ever develop breast cancer, I had two young daughters (ages 10 and 7 at the time). If I carried a BRCA mutation, each of them would be at 50 percent risk of inheriting it from me and, if one did, she would have up to a 90 percent risk of developing breast cancer during her adult life.[28]

I was lucky. My genome sequence revealed I don't carry the Ashkenazi breast/ovarian cancer mutations or, as far as I can tell, any other BRCA mutations. But if I did carry one, would I have disclosed it?

If you'd asked five years ago, I would have said no way—indeed, I made life difficult for Church and other PGP scientists in 2007 when I made this preference known: I wanted to redact my BRCA sequence. Both Steven Pinker and James Watson opted not to find out their genetic Alzheimer's risk when they were among the first to have their genomes sequenced.[29] But even before I got my results, I began to feel increasingly hypocritical about the prospect of redacting my BRCA genes or anything else. I'm all for preserving the Right Not To Know, but it seems to me that if you're at the head of the line, like I was, saying "I *really, really* want the all-you-can-eat buffet," then you forfeit the right to the à la carte menu.

Today I wouldn't hesitate to say yes to full release with no redaction. I understand why someone at risk for Huntington's disease might not want to share that with the world. But I wonder whether we will ever treat HD patients with the dignity they deserve if they don't confront us and force our hand, just as people with HIV forced the world's hand—and especially the hand of the Food and Drug Administration—in the 1980s and 1990s.

We pick our battles in life, and demystifying genetic information is the one I have chosen. If that means fighting stigmatization, discrimination, insurance companies, the FBI, Cylons, Raëlians, or anyone else who would presume to use my DNA against my family, then that is what I will do.

## Of Robots and Hairballs

Mine is a utopian and unrealistic vision. It is not a panacea. Sharing genetic information, let alone making it public, is not for everyone. Indeed, maybe it's only for a small minority. But we "disclosers" are out here and our numbers are growing. At this writing, the Personal Genome Project is pushing two thousand enrollees, with thousands more in the queue. A Silicon Valley company known as 23andMe, launched in 2007, takes customers' saliva and, for a fee, analyzes several hundred thousand genetic markers, some of which have been associated with various traits and diseases;[30] as of mid-2012 it had more than 150,000 people in its database. PatientsLikeMe.com is a commercial portal where patients with all sorts of maladies gather and compare symptoms, medications, diets, lifestyles, and so on.[31] PLM generates self-reported data with *no guarantees of privacy*, yet it, too, has more than 150,000 members. And a variety of other ways for individuals to manage their own data and participate in research have sprung up outside the usual pathways.[32]

Why is this happening? I suspect much of it is due to demand and opportunity: we are curious about ourselves; we believe in science and medicine but are frustrated with existing research and health care bureaucracies, and now we have the tools to bypass them. More than a billion of us have opted to share all sorts of information about ourselves on Facebook and Google—information that strikes me as much more powerful and potentially harmful than almost anything that might be revealed by DNA. Moreover, we are willing to let Facebook and Google *track* and *monetize* information about our likes, dislikes, successes, lapses, and peccadilloes—information we used to regard as secret.

I remain deeply ambivalent about this. I'm not interested in curating my genome, but I confess to wanting some measure of privacy and control over my online presence, which is by now almost certainly a quixotic goal. (For what it's worth, I am not yet on Facebook.)

I will continue to teach my freshman course for as long as Duke will indulge me. Because DNA is fascinating: as the molecule of heredity, as an arbiter of health, and as a cultural icon. People should know more about it.

But destiny it ain't.

When we indulge in genetic determinism and genetic exceptionalism— whether to prevent history from repeating itself, to instill fear, to justify the status quo, or to convince people to give us money—we sell humanity short. As biologist Arthur Lander has observed,[33] and as recent data about all the wacky, unpredictable stuff going on in the genome attest, the dominant cultural icon for the life sciences should no longer be the double helix. It should be the hairball. We have too many moving parts to be represented by the helix, as beautiful as it is. We are too complicated, too dynamic, too agile, too stochastic, too inscrutable.

So don't patronize us. Only some of what we are is genetic and that part will not enslave us. We see ourselves neither as lumbering Darwinian robots nor as somnambulant disciples. We are thinking, feeling, curious, dynamic human beings.

*Projecting trait and trace*, yes. But contingent. And awake.

# SEVEN

# The Power of Two

*Two Sisters, Two Genes, and Two New Chances at Life*

Anabel Stenzel and Isabel Stenzel Byrnes

·································································································

### Anabel

When I contemplate my genetic identity, an image comes to mind: a magnificent blooming tree with three intertwined branches. The tree has been watered by luck, compassion, and resilience.

The first branch is made of the genes of my biracial heritage. My Japanese mother, Hatsuko, and German father, Reiner, both raised in war-torn countries, came to America for higher education. They met in Southern California and fell in love in the mid-1960s. In their genes, they carried core values from their cultures. From my mother's Japanese heritage came humility, discipline, stoicism, and most of all, *gaman*, the Japanese word for "perseverance." My father, a German physicist, passed down to his children traits of curiosity, skepticism, the love of nature, and pursuit of the facts.

My parents shared compatible childhood stories of the Allied Occupation and their adult pursuits of the American Dream. Complementing each other's personalities in the most unusual ways, they wed in 1967. In 1970, our mother gave birth to a healthy boy named Ryuta, meaning "Big Dragon" in Japanese. Nine months later, my mother found herself pregnant again, uncertain about her ability to raise another baby so soon. After all, her own family was so far away and she had limited support.

My mother found herself in labor six weeks early. Just two hours prior to delivery, my mother was told that she was carrying twins. A mysterious miracle transpired when the zygote split and formed two identical twins—and the second branch of our genetic identity began to sprout. Our magical bond began in utero and would be reinforced throughout our lives. On January 8, 1972, I was born first, and two minutes later, twin B arrived. We were named Anabel Mariko and Isabel Yuriko.

Unbeknownst to my parents, they both carried the recessive gene for cystic fibrosis, the most common life-threatening genetic disease in the United States. CF causes an imbalance of salt and water in cells, resulting in thick mucus that clogs the lungs, sinuses, and digestive tract. CF mostly affects Caucasians and is extremely rare in Asians. If two carriers marry, there is a 25 percent chance that each pregnancy will result in a child with CF.

My German father carried the common Northern European CF mutation, called delta F508—a deletion at position 508 of the CF gene. About 70 percent of people with CF carry this mutation. My mother, a pure Japanese woman whose lineage boasted Samurai blood, beat the odds: in 1972, she was told that only 1 in 90,000 individuals of Japanese descent carried the CF gene (this statistic has changed today). In her case, we found out decades later that she carried two CF mutations on one chromosome—R347H and D979A—which may explain her lifetime of minor CF quasi-symptoms such as sinusitis and gastrointestinal distress with fatty foods. One severe mutation, delta F508, coupled with a mild mutation, R347H, would probably have given us an easier life; however, an extra mutation branded us with aggressive disease.

Within days of my birth, my belly swelled like a little watermelon. This was caused by an intestinal obstruction, which occurs in about 15 percent of babies with CF. Because of this strong clue, Isabel and I received a diagnostic test for CF called a sweat test. We were wrapped in plastic and then warmed up. The salt content in our sweat was measured. Our results came back abnormally elevated—positive for CF. At the time, the life expectancy for children with CF was around ten years. My parents were told of our grim prognosis and were, naturally, devastated.

The CF gene in my family, passed down to my identical twin sister and me, became the third and, by far, the strongest branch of my genetic identity. Metaphorically speaking, this branch is covered with thorns as well as flowers. From the challenges of CF, our family quickly learned to embrace an appreciation of life that made us who we are.

With shock, fortitude, and determination, my parents took us home from the hospital six weeks after our birth. They started living out their lives in America with three babies under the age of 2, two of whom required overwhelming medical care. The thick mucus in our lungs formed a breeding ground for infections, leading to chronic pneumonias that would eventually destroy them. My parents were trained to administer inhalation treatments to help loosen the mucus, followed by clapping on our backs as we lay over their laps. These sessions were required three times a day for thirty minutes to help us clear our lungs. We also required frequent antibiotics, vitamins, digestive enzyme pills with every bite of food, as well as a high-calorie diet to counter the malnutrition that came with poor digestion.

Despite some depression and isolation, my parents persevered and maintained a positive, hopeful outlook. They resolved to give us the best life possible, for as long as possible. Amid treatments and hospital stays, we attended school, joined Girl Scouts and a swim team, and traveled to Japan and Germany. My parents watered the tree of our family life in ways that allowed it to flourish. My father's role was to be the breadwinner and the director of "normal" life—organizing family camping trips and encouraging us to exercise to clear our lungs, by hiking and swimming at our local beach. My mother's role was to worry, feed us, oversee our medical care, and keep the family connected. In the forthcoming years, our mother returned to school—first to become a respiratory therapist so she could better administer our care, and later to graduate school to become a medical social worker to help other families with chronically ill children.

### Isabel

As we grew up, Anabel and I quickly formed a symbiotic yet competitive relationship—we shared this disease and were in it together. Our twinship became a tool of survival, and our relationship became a blend of roles of parent, sibling, spouse, therapist, and best friend. If she did her treatments, so must I. We competed by seeing how rigorously we could tolerate our back-pounding treatments and how much mucus we could cough up. We fought miserably if one of us dared to receive less than an equal hand in the burden of our medical care. This behavior continues until today, with most of our arguments revolving around health needs.

A pivotal motivator in our medical adherence came when we were 11 and began to attend an annual summer camp for children with CF. Here, we found a sacred space for our unique culture of CF. All of the campers coughed, were skinny, and drank pills when they ate. Most of all, we all lived with the knowledge that our lives would be shortened. Internally, we knew we had to nurture our spirits and cherish each day. In the next decade, we were exposed to unconditional love, acceptance, and compassion at this camp. Yet we also experienced heartache from the frequent death of friends, knowing that we, too, would someday share that fate. Anabel and I coped with this loss by delving into schoolwork and art. The busier and more distracted we were, the more our minds were spared from thoughts of dying.

Our family life was no different from that of many families with chronically ill children—it was full of tension surrounding family roles, sibling jealousy, and financial challenges. My healthy brother was often left alone while my parents helped us with treatments or visited us in the hospital. Our long hospitalizations created an emotional distance between my brother and the two of us that remains today. My brother's world consisted of Boy Scouts, motorcycles, dating, experimentation with drugs—typical teenage stuff. Our world consisted of illness, hospital jargon, dying friends, and nightmares of suffocation. While we envied him, he envied the attention we received from medical providers, teachers, camp counselors, and even the Make-A-Wish Foundation. Looking back, I still don't know what he truly felt for all those years, but I imagine he wondered why he was spared from having this genetic disease.

Despite my parents' dedication to us, their vastly different cultures created inevitable marital division. Over the years, our illness progressed and required more frequent hospitalizations and more diligent around-the-clock treatments. This stress drove a wedge between my parents. Soon, carrying the CF gene became one of the only things they had in common; yet they vowed to stay together to fight this shared challenge—no matter what. My father plunged into work as his distraction, carrying the burden of providing health insurance. My mother became resentful, exhausted, and depressed by her care-giving needs. Perhaps underneath these dynamics lay the inevitable guilt over the suffering their genes caused their daughters. Yet, my parents coped by being rational: they didn't know they carried the CF gene, so why should they feel guilty about passing it on to us? Guilt was shoved under the rug so that all of our energies could be devoted to the day-to-day tasks of living with cystic fibrosis.

Anabel and I knew that our illness burdened our parents. Our heavy coughing, lung bleeds, and urgent hospitalizations clearly made everyone miserable. Ours was a family that didn't communicate well about emotions—so we simply existed. An unspoken need to compensate for our damaged bodies drove us to become "pleasers." We never rebelled against our parents by disobeying them. Instead, throughout our lives, we tried to be easy, high-achieving kids that any parent would want. My mother held deeply engrained Japanese views toward illness. Double-paned windows to shield the neighbors from our coughing noise and my mother's persistent nagging to hide all our medicine bottles when guests arrived were clear evidence of the shame and stigma she felt about our illness. To those around us, it was simply easier to smile and act like everything was normal. Outside of the hospital, we lived privately with CF, and neighbors and relatives hardly knew what difficulty this illness imposed on our family every day.

Malnutrition and lung disease kept our bodies from growing, so by the age of 18, we looked 12. But, we focused on the positive. Without puberty and hormonal distractions, we concentrated on more important things and didn't get caught up in all the teenage angst that plagued our peers. Being different made its mark on our fragile self-esteem, but we channeled our energies into academics and entrepreneurial endeavors—painting (and selling) T-shirts for nurses while hospitalized and writing (and selling) a booklet about our hospital stays, to educate others. Twenty-plus hospital stays later, Anabel and I found ourselves graduating from high school as valedictorians and being accepted into Stanford University. Yet, our proudest feat was the fact that we were still alive. Thanks to an accommodating educational system that granted equal opportunity to youth with special needs, we *could* graduate high school and go to college.

I was a twin, but I craved my independence. Yet we knew our survival in college depended on our ability to help each other with treatments. Sharing a dorm room, we became closer than most twins. When Anabel started a sentence, I completed it. On our own, we sprouted, and our symbiosis helped this new tree grow. Though close, we competed ferociously at times, argued often, and desperately sought out our own identities outside of being "the sick twins."

In college, our new cystic fibrosis specialists in Northern California introduced us to the most current CF medical care and clinical trials. Thanks to recombinant DNA technology, we entered a clinical trial for DNAse, an enzyme that reduces the viscosity of CF mucus in the lungs. The drug gave us a 20 percent increase in lung capacity and a thirty-pound weight gain (and puberty!). Our renewed health allowed us to flourish in college, finally blending in with our "normal" peers.

The local CF organization, Cystic Fibrosis Research Inc., CFRI, exposed us to many nurturing and inspiring families near Stanford who were living with cystic fibrosis. We met adult CF mentors who taught us how to truly thrive with CF. This community provided us with invaluable tools: patient advocacy, medical knowledge, a drive to be compliant, humor, emotional expression, and fellowship. Finally, I arrived at a place of acceptance. CF became integral to my identity. I could not deny who I was; I could not fight my fate. And I would no longer hide the truth or pretend I was normal. CF was part of our lives, and part of who we were. By the age of 21, we joined the distinguished ranks of being CF adults, a group of survivors whom I have found to possess unique character traits of obstinacy, heightened body awareness, vigilant self-care, humor, and feistiness.

During our senior year at Stanford, I met Andrew, a charming, romantic Irish American intellectual. My venture into dating came with insecurities typical of anyone with an illness. Why would anyone want to be in a relationship with someone sick? Let's face it—CF isn't sexy. How could mucus, constipation, and flatulence be attractive? But, illness tests the strength of human love, and over the years, even long-distance, Andrew proved his acceptance of CF and all things involved. As soon as the doctor found out about my serious boyfriend, he gave me "the talk" about birth control. "With your lung capacity, it's not safe to carry a pregnancy," he warned. At 22, both Andrew and I laughed off his advice, as we had no interest in parenthood. We were in a selfish period, focused on career goals and just being in love. And as we matured, Andrew's increasing lack of interest in children consoled my fears that being childless would be too much to ask of him.

## Anabel

After college, Isabel and I were healthy enough to live in Southern Japan for a year, to teach English. We were cared for by a Japanese pulmonologist who was fascinated by CF. When my grandmother visited us from Tokyo, the doctor tested her blood for the CF gene. The CF gene had been discovered in 1989 by Francis Collins and Lap-Chee Tsui; by 1995, more than a thousand mutations had been found in the CF gene that could cause CF. A month later, we were told that my grandmother was not a CF carrier, but my mother carried two CF mutations on one chromosome. "I knew all along. It was from your grandfather's side," my grandmother insisted. My grandfather came from a large family of eleven children, seven of whom had died in childhood. He himself had died of pneumonia in a prisoner-of-war camp during World War II. We will never know if any of our Japanese relatives suffered from cystic fibrosis. We can only speculate that the double mutant allele could have contributed to their weakened predispositions.

After a year in Japan, Isabel and I were accepted into graduate school at the University of California, Berkeley. Like my mother, Isabel pursued a career in medical social work, while I pursued a career in genetic counseling. We both were routinely annoyed by much ignorance about CF from our past hospitalizations, and we wished to join the medical profession so we could educate others about living with chronic illnesses like CF. While our declining health in graduate school posed a constant challenge, the advent of home intravenous antibiotic therapy spared us from frequent hospitalizations. With hard-headedness, Isabel and I were able to complete master's degrees. Our obsessive time management, coupled with sleep deprivation and a limited social life, allowed us to blend in with our classmates while completing four to five hours of medical treatments daily in the privacy of our home.

I began my career as a prenatal genetic counselor. Through a rigorous training program, I learned to counsel families facing many genetic disorders, including CF. By the late 1990s, carrier testing and prenatal diagnosis for CF were routine—something my parents could never have imagined. Through my own professional decision not to disclose my illness, I was able to maintain an objective yet compassionate approach with families facing a prenatal CF diagnosis. These families were making deeply personal decisions to continue or terminate their pregnancies, and knowing that I had CF would only introduce bias. My 30 percent lung capacity was made evident by my pale face and emaciated posture, but only those truly familiar with the "CF look" could possibly guess that I had it. Most of my patients were immersed in their own genetic anxiety and never suspected that I was sitting in front of them as an example of what they could face if they continued their pregnancy. I recall feeling shameful and dishonest when a biracial couple declined CF carrier testing, saying, "Oh, we're mixed race. We don't have to worry about CF." Or, in another scenario, a couple whose pregnancy was affected by CF said, "We'd like to talk to a doctor who specializes in CF. They'd probably know more about CF than you do." But in these situations, keeping my silence was the right thing to do. I met several of these couples years later at a CF conference. Then, I disclosed that I had CF and provided an apologetic explanation of my professional ethics. They responded only with intrigue, admiration, and wide eyes of hope, inspired that their children, too, might grow up and thrive despite having CF.

Around the time that my career began, my lungs started to fail. I was heavily in student debt as CF slowly took over my life, allowing me to work only four hours per day. Thankfully, Stanford Hospital was a nationally recognized lung transplant center, and I was evaluated for a double lung transplant. Lungs are matched by size and blood type only—not by tissue typing for a genetic match. At 24, I was listed for a double lung transplant, modern medicine's last-ditch effort to save people with CF. After all these years of fighting CF, despite my best efforts, CF was winning. I couldn't last long enough for the promised cure that we hoped for when the CF gene was discovered decades earlier. Without a cure, magic pill, or gene therapy to stop the ruthless downward spiral of CF, I was reaching the end of my life.

The journey to accepting a lung transplant was not easy—the illusion that I could control CF by being a compliant patient evaporated, and I plunged into a sense of failure. Why was I the sicker twin despite having the same genes and environment as Isabel? We were a symbiotic team fighting a parallel battle throughout our lives, yet somehow I was less fortunate. Just as with boyfriends, grades, and even job opportunities, Isabel always got the better deal.

After a rigorous evaluation process, I was placed on the national waiting list, kicking and screaming. It was all so unfair. CF used to be a disease that killed us in childhood, before we could imagine any dreams; now it was a disease that killed us in young adulthood as we pursued our dreams, tempted by a normal life of careers, relationships, and existential maturity.

As the lung transplant became a reality, I focused on the life I wished for—to be free from CF. I made a list of things I wanted to do—swim, run, hike mountains, travel. And yet I thought of my donor—somewhere out there he or she was living, likely to die soon, tragically. I struggled with survival guilt—someone would have to die for me to live. Was I worthy of that?

After sixteen months on the waiting list, on June 14, 2000, I received the call telling me that donor lungs were available. My life had become one big struggle to breathe and I was exhausted. I was finally in a state of acceptance, ready to take on a new journey, or at least die trying, with this high-risk surgery. Surgery took nine hours. I was hospitalized for only twelve days—the shortest hospitalization of my life! I later learned that my donor was a 29-year-old man from Oregon who had suffered a brain aneurysm. Despite every effort to save him, he became brain dead, with no hope of survival. His family said yes to organ donation at a time of such tragedy so that five other strangers could live. This is the ultimate gift of humanity.

I awoke from my surgery with a strong awareness to live fully for my donor. Within five months of my surgery, I had normal lung capacity and had more energy and breath than I had experienced my entire life. I was like a kid in a candy store, let loose in the sweetness of health and opportunity. I immediately launched into a life of volunteer work, my career, and active hobbies like hiking, backpacking, jogging, and swimming. Finally, I was free from hours of medical treatments and constant coughing, congestion, and fatigue. Though I had to take daily immunosuppressive medication to prevent my body from rejecting the donor's lungs, my German and Japanese genes granted me the strong pharmacogenetic ability to process these drugs with minimal side effects. My tree was in full bloom—springtime had arrived.

## Isabel

While Anabel was caring for pregnant women, I became a pediatric social worker, working just down the hall at the same children's hospital, in
the Infant Development Clinic. I was caring for families with babies who
had spent time in the Neonatal Intensive Care Unit. Ironically, some of
those babies were born to parents Anabel had counseled prenatally—
thereby thoroughly confusing them! In this setting, I, too, hid my CF and
tried very hard to present myself as completely normal. Inside, though, I
knew my illness gave me enhanced empathy and a network of resources
and tools to be an effective social worker.

Though I was always stronger than Anabel, my deterioration was
just a few steps behind hers. Like my sister, I admit that I pushed myself
to the point of masochism to keep working. On several occasions, despite
a flare-up of pneumonia, I resisted my body's warning signs because I
didn't want to miss work and inconvenience my co-workers. Ultimately,
my body would break down and a crisis would force me to surrender,
going into the hospital in a wheelchair, gasping for breath, or through the
emergency room, with bleeding lungs because I had waited too long. The
drive to be normal proved to be detrimental to my health and probably
hastened my decline. Looking back, what I needed was a hard bonk on
the head to tell me to stop.

Three years after Anabel's successful lung transplant, it was my turn.
By 32, I had faced several life-threatening lung bleeds, had become
oxygen dependent, and was forced to quit my job for a more stable life
on disability. Soon, I, too, was reluctantly walking through the doors of
the Stanford Lung Transplant clinic for my evaluation. Thankfully, I was
deemed an excellent candidate, but I didn't feel ready.

At that time, I decided to use my free time to review my life and start
to write a memoir. With Anabel's blessing it would be a twin memoir. As
my body weakened and my mind raced, I took writing classes and joined
writing groups. I assigned Anabel various chapters, and we took turns
recording events in our lives. My mother had encouraged us to journal as
children during our hospitalizations, so we had volumes of written material for inspiration. Writing became an immensely cathartic and healing
experience as we reflected on our illness, our family, and our twin dynamics. Within eighteen months, we had written more than 350 pages.

Midway through the writing process, in January 2004, I was hospitalized. I became unresponsive to the usual antibiotics, and my condition rapidly deteriorated. My husband and Anabel stood helplessly by my bedside, coming to the realization that I was dying. Though I was actively listed for a lung transplant, I was not at the top of the list and the prospects for a transplant were slim. I can only imagine the despair that Anabel must have felt, witnessing my end of life—from a shared disease. I would not have the same opportunity of experiencing life with new lungs that she'd had, and after years of her being envious of me, this would serve as a sort of twisted divine justice. Anabel and I had come into the world together and always knew that one of us would die first; but she wasn't ready to be a twinless twin. Andrew and Anabel, as well as my parents and friends, prayed for a miracle but prepared for a funeral. My lungs failed and I was put on a ventilator, a last-ditch effort to hold on in the hope that donor lungs would become available. Just before I lost consciousness, I spoke in a near-coma delirium, insisting, "There's going to be a miracle."

After a nail-biting twenty-three hours that would put TV medical dramas to shame, my family heard the most beautiful words, which dropped Andrew to his knees: "We have lungs for Isabel." At the eleventh hour, with an estimated twelve hours left to live, I received the gift of life.

I learned later than my donor was an 18-year-old Mexican American man from Central California. Xavier Cervantes died from a brain injury sustained in a car accident. Two months before his car accident, Xavier had told his mother that if anything ever happened to him, he wanted to be an organ donor so that he could help people. Xavier is our savior: this young man infused life back into me and brought me back to be with my twin and my husband. It was a divine gift, a spiritual alignment of stars that occurred at the perfect moment.

For the first time in our lives, Anabel and I breathed through different lungs—lungs that were genetically distinct from our own. These lungs would never have the CF gene in them, ever again. Were we still twins? It didn't really matter, because within months, we were exploding with our new life, free of CF, exercising passionately and pursuing our book publication. The doctor told my parents, "Congratulations, your children will not die of CF lung disease." Who would ever have imagined? My mother prayed to God; our father thanked modern science for the miracle of transplantation. This was truly their American Dream.

## Anabel

Our book was finally complete, and by the fall of 2006, we had secured a publisher. The book became our focus.

Just about that time, however, my body began to reject my lung transplant, for reasons we don't understand. Doctors said, "All lung transplants have an expiration date." Lung transplants are the most vulnerable organ transplants, with only 50 percent of recipients surviving five years. After six fabulous years, once again I struggled to breathe. Within eight months, I was back on oxygen and in a wheelchair. Meanwhile, like an imbalanced scale, Isabel thrived with her new lungs. She learned to play the bagpipes, hiked the Grand Canyon, and ran three half-marathons. Repeating our past, she was the strong one again. But she also carried the heaviness that her party, too, would be temporary. Worse, she asked me if her life was worth enjoying if I died. I prepared for death, accepting that a lung transplant is life-extending, not life-saving. I was grateful for the time I had and prayed to those friends I had lost for courage in the dying process.

By the grace of God, the goodness of the medical team, and medical insurance, I was offered a second lung transplant. At first, I felt guilty. Some of my CF friends had died waiting for their first transplants. Who was I to deserve another donor lung, which was such a limited resource? After much soul searching and Isabel's encouragement, I realized that God was opening this door for me, and someday, all the doors would be closed. So, I was going to walk through this one. Besides, with the book, I was too busy to die!

After three months on the waiting list, I received another lung transplant, at age 35. The second transplant surgery was precarious and painful. I joined the privileged few dozen people in the country to survive a second double lung transplant. After four months of rehabilitation, I decided I was sick and tired of being sick and went on with my life. Our memoir was on the verge of being released, and it was time to get this show on the road.

The publication of our memoir, *The Power of Two: A Twin Triumph over Cystic Fibrosis*, in 2007, launched us on several national book tours. Perhaps it was our humorous twin dynamic, the depth of our lectures, or our enthusiastic energy that seemed compelling to audiences, but soon we were bombarded with invitations to speak at conferences, medical schools, pharmaceutical companies, and community gatherings. We spoke both personally and professionally about the struggles and blessings of living with genetic disease, while sharing the wisdom learned from a lifetime of illness and proximity to death.

In our healthy, active state, we are often asked if we want children. It's strange how people push the envelope of what's normal for women our age! Despite both being married, we have chosen never to have our own children. Our lungs and CF guts *are* our children—they require 24/7 care and are troublesome enough. We cannot afford the time and energy to take care of children. We both firmly believe in a "no child left behind" policy when living with a life-threatening condition, as we witnessed the psychological effects of parental loss in my mother and others. For that reason, we also choose not to adopt or have a child through surrogacy. Fortunately, we both found husbands who aren't the parental types, and it wouldn't be fair to burden these reluctant men with single fatherhood if—no, when—we died. Unlike some of our CF peers, we also were lucky that our in-laws were accepting of our choices and not the types to hold grudges or incessantly insist on grandchildren. We had to accept the fact that motherhood would be reserved for another lifetime, and that was okay. On several occasions, we have received "the look" from healthy women in our midst who pity our childless lives. However, we have sought to view motherhood as another privileged journey in life that is not for everyone. Freed from child care, Isabel and I are able to pursue many more activities and passions that most of our peers who are raising children cannot.

And, there will be no passing on of the CF gene from our DNA. With my husband being part African American, his chances of being a CF carrier are reduced. Isabel's Irish American husband was actually tested, just for fun, and found not to be a CF carrier. So her risk of passing on CF to her children is nearly zero. I have to think hard why my attitude toward not having children is so nonchalant. Perhaps it was the trauma of hearing babies crying day in and day out while hospitalized as a youth. I had no desire to ever hear that unpleasant screaming sound in my own home. Isabel, on the other hand, sometimes expresses her sadness about being childless. She often talks about her fantasy of what her child would look like, or how she'd raise him or her. However, I focus on the gift of still being alive—how can we dare ask for more?

So, our book became a form of creative procreation—passing on our own legacy and our own "genetic heritage." Our passion was and still is to raise awareness about cystic fibrosis and to share the benefits of organ donation. We hope to move people to sign up to become organ donors so that others may receive the gift of life that we have. We also hope to reduce the mystery and stigma about living with genetic diseases like CF, with an added intention that more research dollars will be raised to find cures.

## Isabel

Serendipity has made our lives colorful. In 2009, after my mother shared our book with a long-time friend who was also a Japanese editor, a prominent publisher in Japan decided to translate and publish our book. Organ donation remains controversial in Japan. In July 2009, the Japanese Parliament had just passed the first law in twenty years that would ease restrictions on organ donation. The publisher felt that a personal account by organ recipients would offer valuable education to the Japanese public.

Together with supporters, Anabel and I organized a twenty-six-day, ten-city book tour across Japan, speaking at medical institutions, public lectures, and schools about organ donation. In general, Japanese culture has strong religious taboos against handling a deceased body and also questions the notion of brain death, which feeds mistrust in the transplant process. Also, in Japan, giving to others is based on reciprocity; giving to strangers is not customary. Therefore, donating organs to a stranger baffles many Japanese people. As we learned growing up, illness is stigmatized in Japan, and patients are often isolated and private about their conditions, for fear of ostracism. Our speeches, given in Japanese with tremendous effort, touched audiences and, hopefully, opened hearts and minds to the perspective of patients. Since Anabel and I both had the rare chance to meet our donor families, we also emphasized the value of organ donation for the donor families. By providing life to others through their loved one's death, some of their grief was eased. This adventure was awesome, but through it all, our twinship was challenged. We bickered intensely, resenting each other for trying to control the content of our speeches. We faced the pressure of speaking formal Japanese, a language in which we are far from fluent. Amazingly, our bodies withstood the rigors of this tour.

During our travels, we were hosted by the Committee to Enable CF Treatment, a brand new advocacy group organized by CF families in Japan. We had always known there were a handful of CF kids in Japan. We had heard tragic stories of their early deaths, lack of any CF care, and near impossible quests for lung transplants. The average life expectancy of CF patients in Japan was fifteen years, because they did not have access to any of the CF medications available in the United States and Europe. This disease was simply too rare for companies to profit from distribution of these drugs. Before our trip, we asked our American peers for any unneeded medications, and we donated large quantities to share with the Japanese CF patients, on humanitarian grounds.

Anabel and I were overcome with emotions during our first meeting with these CF families. Finally, after spending our entire lives feeling as if we were some genetic anomaly, we could meet other Japanese people living with our disease. We cried over the desperation of these undertreated CF patients in this rich, advanced country. We were reminded of how our own fate could have been so different if we had been born in Japan. The randomness of opportunity seemed so unfair. We met with parents whose children had died, and those trying to save their kids, and we even met three adorable CF kids. We shared stories of struggle, support, and perseverance. Interestingly, two of the parents came from the same hometown as our grandfather.

In another case of serendipity, my husband met a filmmaker interested in social causes and in using film as an agent for change. The two men teamed up, raised funds. and brought a documentary film crew with us to record our advocacy efforts and the organ donation situation in Japan. Academy Award–nominated director Marc Smolowitz embarked on a two-year journey of raising money, collecting 240 hours of footage, and traveling across Japan and the United States for film shoots. After five months of intense editing, his inspiring film, *The Power Of Two*, was released in August 2011.[1] Anabel and I became centerpiece characters of the film as examples of patients-turned-advocates for the cause of CF and organ donation. Today, the film has been viewed internationally at film festivals, at community screenings, with the DVD release, and through our distributors and has received more than ten awards. It was released in theaters in Japan in November 2012, after debuting at the prestigious Tokyo International Film Festival in 2011. This overwhelming public portrayal of our lives, and the positive response, reminds us that life with cystic fibrosis is full of extraordinary surprises.

## Anabel

In 2012, Isabel and I had the privilege of celebrating our fortieth birthday. After more than sixty hospitalizations and three lung transplants, we have managed to live four times longer than our prognosis at birth. Our lives have been made possible by the confluence of factors—America's excellent medical care, the grace of God, wonderfully supportive parents, and a nurturing, loving illness community. I have lived long enough to buy a home and get married, in 2010. Isabel and I continue to grow in our careers. We have welcomed the birth of two beautiful nieces to my brother and his wife. They are three-quarters Japanese and are healthy, since, thankfully, through genetic testing, my brother was found not to be a CF carrier.

The three branches of our splendid family tree—our culture, our twinship, and yes, even our CF genes—have empowered us to do more, be more, and grab more out of life than most people our age. Every branch of our tree has absorbed as much oxygen and sunshine as possible, radiating eagerness and gratitude. Though our tree has weathered many seasons, with each new season we see different shades of nature unfolding. Though fully aware that we may someday enter another season, this only teaches us to live harder, love deeper, and breathe in our miracles with greater joy.

# EIGHT

# Permission to Look

## Documenting the BRCA Mutation

Joanna Rudnick

...........................................................................

EDITOR'S NOTE: *In 2003—two years after learning she was positive for the BRCA1 mutation, at the age of 27—science journalist and filmmaker Joanna Rudnick began work on a documentary situating her experience in a wider study of the mutation's social and psychological consequences (*In the Family, Kartemquin Films, Inthefamilyfilm.com)*. The film is built around video journals exploring Joanna's unfolding sense, in her early thirties, of what the mutation meant, in terms of both personal relationships (with her then-boyfriend Jimmy, her BRCA-negative older sister, her parents, doctors, and friends) and decisions about work, relationships, and childbearing. Joanna alternates these video journals with interviews of people of different ages, ethnicity, and socioeconomic positions, all grappling with the mutation: Linda Pedraza, a woman with end-stage hereditary cancer, and her husband Luis; Martha Haley, a woman from a high-risk family unable to afford BRCA testing; the Hanke sisters, who learn the results of BRCA testing in the office of Mary-Claire King, the scientist who discovered the gene in 1990; and Mark Skolnick, founder and chief scientific officer of Myriad Genetics, the Utah-based firm that holds patents on the BRCA genes and its known mutations.*

*Initially, Joanna didn't plan to reveal her genetic status in the film. Partly, the decision to ground the documentary in her own experience came from necessity: "I was looking for a character like me during the pre-interview process, but at the time, could not find a mutation-positive, single woman who had not [yet] had surgery and was willing to go on the record, on camera with her story," she notes. Exposing her own story proved crucial to the film, where personal aspects of living with a BRCA mutation intersect with larger social and ethical issues.*

*I first saw* In the Family *several years ago, when I was researching the history of gene patents at stake in the lawsuit between Myriad Genetics and the ACLU. The video clip of Joanna's exchange with Mark Skolnick led me to the documentary itself. While this material is familiar to me—this is the mutation my family has, after all—I was moved, unsettled, and surprised by the film: by Joanna's candor; by the moments she is willing to reveal in the film, from visits to her gynecologist for transvaginal screening of her ovaries to poignant conversations with Jimmy, her then-boyfriend, as their relationship began to unravel (largely around issues related to BRCA). The honesty of other women interviewed in the film also surprised me; several scenes were difficult for me to watch.*

*Thinking about how the film uses tactics of familiarization and de-familiarization to present these stories, I asked Joanna if she would be willing to talk with me early in August 2012. She was getting ready to move from Los Angeles to San Francisco and was eight months pregnant with her second child (her first, a daughter, was almost 2). I felt, as we talked, as if we could somehow be related. Here we are, the two of us: women from families deeply affected by the BRCA mutation, both committed to bringing stories about the BRCA mutation out in the open, though in different aesthetic forms. We each believe there's much to be gained from grounding discussions of genetic identity in personal experience. I thought asking Joanna about this was a good place to start.*

AMY: What made you decide to make a film about the BRCA mutation? Can you talk about what made you decide to include your own story in the documentary and about the risks and costs of that exposure?

JOANNA: I was in so much denial when I started. I'd found out at 27 I was BRCA-positive and didn't want to dive in right away and start talking about interventions. I didn't really feel like telling all my friends or having long, emotional conversations about the mutation, I didn't want to be identified by it. I moved to Chicago to be close to family, which had a lot to do with learning about BRCA—I wanted to be rooted again. While I was in Chicago, I started to research the mutation, but very much as a dispassionate exercise. The first serious phone call I made [about BRCA] was to Sue Friedman [founder and director of FORCE, a nonprofit for individuals and families at risk for hereditary breast and ovarian cancers]. I told her I wanted to make a film about BRCA, but at that point I didn't tell her I have the mutation. I wasn't conceiving of being in the film then; this was going to be about other women who had the mutation.

When I approached Gordon Quinn at Kartemquin Films—he became the film's executive producer—he said, "You have to be in it." At that point I'd already interviewed a number of women with the mutation, but I hadn't revealed to any of them that I have it, too. That's how locked up I was.

Another reason I put myself in [the film] was because back then, I couldn't find a young woman who was single and BRCA-positive who hadn't had surgery yet. I asked, and looked around, and I wasn't finding that person. Now I would—it's a different landscape than it was in 2003. But not back then. So it was a constant back-and-forth, with Gordon and my colleagues pushing me deeper into the story, encouraging me to reveal more about myself, if I was asking other people to do that. That's where those video journals started coming in.

As for the risks of that exposure . . . When I decided to be in the film, I had a conversation with my family, which I revisit in the film. At first my father was reluctant to encourage me to be in it—he was worried about the personal costs. I worried that I'd lose my anonymity, that I'd be exposed and identified by this mutation, something I was desperately avoiding. I was so locked up, and to appear on public television, in front of everyone . . . in other countries . . . That was an incredibly long process in my evolution. My mom . . . at the time, she was supportive. But to back up, we never talked about my mother's cancer. My mother was diagnosed with ovarian cancer when I was 13, and it was one of those situations where she didn't want everyone talking about it, she wore a wig, she was dealing with it, but it wasn't her whole life, and once she (thankfully, luckily) survived it, we didn't revisit it. So for me to come to her and say, "I'm making this film, and I'm going to talk about cancer, and our family, and our legacy, and I want you to be in it and to talk about it," that really went against the ethos of how my family had dealt with the disease. (My mother's aunt had died of ovarian cancer and her grandmother, of breast cancer, the year before my mother was born, and she was named after both of them, but didn't realize that [they] had died of cancer until she was an adult with her own family.)

So part of the filmmaking was a personal, psychological, almost a spiritual journey that allowed me to go back and cope with a very painful legacy.

My mother is very funny . . . I'd call her early in the morning and ask her questions and she'd say, "Do we really have to talk about cancer before I've had my coffee?"

But I had to ask questions. As a filmmaker, this was the only vehicle I knew in order to come to this subject. I think people really identify with how difficult it is to have these conversations inside of families. Modeling these conversations—whether in a genetic testing session, where you're actually getting and revealing your results, or private conversations with a boyfriend or with a family member . . . I think people relate to overcoming some of these barriers.

AMY: I'm thinking about your interview at Myriad and also that scene with Mary-Claire King and the three [Hanke] sisters who learn at the same time who's positive and who isn't. These two interviews both seem, in different ways, interventions in the knowledge of where we were at the time with BRCA. Even now, we're still fumbling through these things . . . In your interviews, you navigate between the personal, like the scenes with [your then-boyfriend] Jimmy, which seem as intimate as you can get, out to larger legal, social, and ethical issues . . . Which scenes were the most salient or provocative for you?

JOANNA: Myriad's an interesting component of this. One thing that shocked me when I started researching BRCA is that the genes are patented. Both BRCA1 and 2. I didn't know you could patent a gene, and a lot of people have the same response I did. From the informational perspective, I wondered how I was going to stay true to the narrative of the film, and to the characters, and root the audience in their emotional journey, and still go into the story of Myriad. Those were complicated editorial decisions— how much information do you give, how far do you go into these ethical and legal issues without taking audiences away from the narrative of the film? And also, when does the filmmaker turn from being an artist and storyteller to becoming an activist as well?—this wasn't part of my master plan for telling this story.

One of the ways we get to Myriad in the film was through the story of Martha Haley [a woman unable to afford BRCA testing]. I learned that many people couldn't get genetic testing because they couldn't afford it (and at the time, very few people qualified for [Myriad's] special assistance program), and because Myriad has a monopoly on the test and has price-hiked it, up, up, up, and has a vested interest in keeping the price high. There are also insurance complications in getting approved for the test. I knew I had to have a conversation with Mark Skolnick [founder and chief scientific officer of Myriad Genetics].

Strategically, we held off that interview until almost the end of the filmmaking process, it was one of the last interviews we did. I'd gotten to know the women and the issues and I was really grounded in the story before I went there. So when I got to Myriad I was *so* passionate . . . I was there as a journalist, and I was able to ask the questions, but by that point, I knew the stakes and the consequences.

And they were gracious. They let me in—that wouldn't happen today, I know that—it was a very different time then. I saw that it was a really good lab, but at the same time, I saw—and Skolnick himself admits in the interview—that there's no reason the cost of the [BRCA] tests should be that high.

Mary-Claire King was a personal hero—I wanted to interview her from the beginning, to understand how she came to devote so much of her career and life to BRCA. To see a scientist as dedicated as she is actually unveil—that's not the right word—*convey* genetic results to this [Hanke] family. Their conversation goes to the heart of what the film is about: It's not about the individual, it's about the family. In that scene, you get to see three sisters, all deeply impacted by their mother's cancer, deciding they're going to get these results . . . To bring the audience into that moment—I couldn't think of anything more intimate, the stakes were unbelievably high, this family was so close, and so incredible (I just saw them recently, all three sisters together, and they're phenomenal)—with the woman who made this all possible—it *still* makes the hairs stand up on my arms. It's a gift, to see Mary-Claire King disclose those results.

That was one of the more emotional shoots for me. I was nervous. I remember sitting under a table in the corner because I didn't want to be in their sight-line, I knew how difficult that day was for them, and we had to have a camera and sound operator there, but I didn't want any of them to see me, I wanted to stay as far away as I possibly could.

Dr. King and her lab's genetic counselor, Jessica Mandell, had such a positive way to handle getting those results, a dignified way . . . You say we're fumbling through this, and it's true, it's never easy to tell anyone they have this mutation, especially when you're saying, *You have it, you have it, you don't* . . . It's a complicated and delicate dance, of how you do that. I myself had a negative experience with genetic counseling, and though I'm the biggest supporter of it, in my own session, it felt like they went right to, *You have this mutation*, to *This is what you can do, this is how your life is going to be impacted*, to having to deal with all of this right away. I had a very different experience, so I felt like I was being re-counseled as a proxy through these sisters. That was an incredible moment. I'm the biggest Mary-Claire King fan. I admire her so much, and I feel like she should win the Nobel Prize for that discovery [of the BRCA gene]. If it weren't for her, we would not be having this conversation right now.

AMY: How did the Hanke sisters feel about that exposure? It's such an intimate scene. I like how you describe it, as a gift, but I'm wondering how these three women, now, almost ten years later, feel about their participation.

JOANNA: It's interesting, I'd like to ask them how they feel about it today. It's been a long time since I've seen the film with them, but I remember the first time I screened it for them, I was incredibly nervous, because there's a scene in the car on the way to the counseling session—we almost didn't film that, so we didn't really film it right, our audio is bad because we didn't mic it in the right way. It was a difficult moment, they ended up having this conversation on the way to the clinic about the decision to get tested, and the youngest sister said, "Well, you all kind of just decided, and I'm not sure this is what I want," and I think she felt the same way about the camera being there. As a filmmaker, that's a very difficult position to be in—you never want to feel like you're pushing someone beyond their comfort zone or crossing any line. I think the rest of the family was much more open to the cameras being there. And then she ended up testing positive, and there was a part of me that wondered, how's she going to feel when she sees this? How are they all going to feel when they see how honest this moment is? But later, when I screened it for them, they all loved it—they laughed at themselves in the car, they said, "You got us!" I don't think they have any regrets about being in the film. They're real advocates, those women, all in different ways, they've been unbelievably strong, they're proud of being part of the film. In the project I'm working on now, on Rick Guidotti [a photographer whose organization, Positive Exposure, is dedicated to making visible and celebrating genetic difference], I'm finding the same thing—people reveal so much, and you always feel responsible as a filmmaker, you want to be sure they're comfortable with that degree of revelation.

Those sisters were so honest. I hope people relate to the film because of that honesty. Those moments are just so hard, when people leave the doctor's office . . . A lot of doctors and genetic counselors have said to me, "We never *see* this stuff. We don't hear the conversations that happen before you come in or after you leave. This is what we don't see."

In the film, we wanted to take the audience to those moments. In the first proposal I wrote for *In the Family*, describing the results of my own BRCA test, I remember saying, "This is not just a sheet of paper with information on it—*'positive for a deleterious mutation.'* This is everything that happens before and after you get that result, and how transformative that information is on the rest of your life."

And I thought, how better to tell that than to turn the camera on people and let the audience be there with them. I don't know any other way to do that. And in doing that, it allowed me to go through my own process of everything I'd held off.

AMY: You talk about collaborations—with your production team, your editors, people you interview. It strikes me that much of the power of the film for you also comes through collaborations with audiences. There seems to be an "aliveness" for you in these screenings. I know you've been a speaker at some of them, and I wonder if you've been surprised by the responses you've gotten.

JOANNA: I'm always struck by the emotion. When the film ends, there's usually silence. A lot of tears—a lot of women *and* men crying. A lot of healing . . . processing. People have asked me incredibly personal questions—I put myself out there for that, and audiences are emboldened to ask. There's lots of sharing of personal experiences. One common response is also anger over the patents held on BRCA. People feel very cheated that could happen, and they feel shocked and betrayed that research on the genes can be impeded by these patents. They're outraged by that . . . (Of course I have a personal opinion about that as well.) One of the responses that's the most complicated . . . I always thought it was ideal that in the film, I was a character who hadn't made any decisions. I never wanted to model any one way of thinking about intervention, in the physical sense, and that's sometimes a complicated conversation at screenings because people can be very invested in the choices they've made—or that a mother, sister, daughter, or friend has made. That can come up as, "Why wouldn't you do *this*?" But in the film, I really wanted everyone to have a voice—including people without any family history or connection to this mutation whatsoever; people who'd been through cancer and made their treatment options based on that—I wanted everyone to be part of the conversation. People sometimes asked, "Do you really need another character?" or as a team we debated if we should drop a story for timing purposes, but I think everyone in the film is crucial. It's important that we were crossing racial, ethnic, socioeconomic, and age boundaries, that we were challenging stereotypes to find the common denominators; all of that was purposeful. People are so different and have such different life experiences, but they can have such similar responses to this mutation.

One of the interviews that still really stands out for me, a moment that I still can't watch without getting choked up, is my interview with Linda Pedraza, one of the first people I talked to. We went to interview Linda when she had terminal cancer, just months before she passed away. This was a family coping with the eventual loss of a mother, wife, daughter, sister, and friend. I spent time with the family both before and after Linda died, and as the filmmaker—there were moments when she looked directly through the camera, and it almost disappeared, and it was just us, as if she were saying, "You have to take this seriously. You need to see what is happening to me." Linda's mother was diagnosed with ovarian cancer at 43, *my mother* was diagnosed at 43, and Linda was diagnosed at 43. That was chilling. I still can't watch that interview without incredible emotion, without thinking, what a gift, to allow us, at that stage, to have access to her family and her children, when she had so little time left, because she really wanted to help other women. It was that simple for her: she wanted to save other women. Such a smart, brave woman. Experiences like that, for me . . . I felt so lucky to have the vehicle of filmmaking, to hope to be able to help women in that way.

I remember feeling an awesome responsibility walking away with that footage, because at that point, I had no broadcaster, I just had a team of people who believed in me. I think I used all the money I had for that one shoot, and coming back to Chicago, I remember saying, *I need to get my grant money, I need to find more women, I need to do this.*

AMY: I have a question that links to the aftermath of the film in a more personal way. You were so young when you made the film, and there have obviously been lots of changes for you. Now you have a daughter, and another on the way, and I'm imagining you see some things differently now. What about the role of the mutation as you've become a mother, as you've moved into your thirties? For many women with BRCA mutations, the issue of time is so vexing . . . the question of a biological clock, and the arc of time in one's life is clearly impacted by this mutation, so I'm wondering if you'd be willing to talk about that.

JOANNA: When I was making and rolling out *In the Family*, I was so entrenched in BRCA that I had to take a personal break from the mutation when I finished the film. Of course, there are no real vacations, and my screenings and fears never fully took a backseat, but they weren't as all-encompassing. This allowed me to meet my husband and have a real relationship.

Now, my BRCA story has come roaring in along with motherhood. I'm 38 now, I'm about to have my second child. The mutation is back as a deeper reality now that I'm much closer to acting on the decisions that I talked about in the film. It's become even scarier as I get closer to the age my mother was when she was diagnosed . . . Having my first daughter, and now my second . . . Of course [wryly] I'm having two girls, with this mutation, you can't help but look at them and wonder, What's life going to be like for them? Did I pass this on? I'm going through all the things mothers in the film were going through. And facing the poignant reality of wanting to be around for them. I'm about to give birth [in September 2012] and six months later, I'm going to have my ovaries out. So I'm less than a year away from surgery—the first surgery. That's deeply emotional for me. I'm grateful that I was able to have my children. I was worried that I wasn't going to be able to get here, and I know there are other single women who are consumed by that same fear. That was a big part of what I felt during the film: How am I going to find the right someone who will accept me like this? Am I going to be able to have children in time? That crazy time clock that's associated with BRCA. That was always the dream, to be a mother, and that was the one thing I was so afraid of the mutation taking away from me. There was a lot of battleground thinking along the lines of "it's taking this from me." Now, I'm grateful, it's empowered me, but that's taken more than ten years. There was real anger for me at the mutation . . . Now, there's more of an acceptance, and a surrender, in a sense, to what I have to do.

Something I didn't put in the film, and now I ask myself why I didn't . . . I always wanted to know more about my mother's aunt, who died before my mother was born of what the family lore identified as a female cancer. We didn't know any of the details—how old she was when she died, or how long she was sick, or if it was actually ovarian cancer. So I went through ancestry.com and found her gravesite . . . I sent away for her death certificate and I got a copy, it's right here on my desk in front of me . . . It came in the mail, and I opened it up, and I was shocked, I never knew this, it turned out she died when she was 37. And I'm 38 now, but I got the certificate right before I turned 38. I opened it up, and it was incredibly detailed; it said *carcinoma of both ovaries*. This was back in the 1940s. She left a very young child behind.

I thought, there's a reason that I got this [document] when I did. I felt like I was punched in the stomach, it was the reality of it, and all of that pressure, and the time . . . There is no time anymore, there's just the plan now. I don't have the luxury of time anymore. And I don't have the luxury of being just one person who doesn't have a family that relies on me.

So everything's changed. It's scary, and in one way gratifying that there's this option [surgery]. I just need to get through . . . it's a lot . . . [laughs]. Yeah, life has changed so much.

I called Luis Pedraza [Linda Pedraza's widower] on my thirty-eighth birthday, because in the film, I'd told him, "I'm going to call you on my thirty-eighth birthday and tell you that I'm having my ovaries out!" I promised him that. So I called him on my birthday and I told him, *I'm pregnant, I'm due in September, but as soon as I have the baby, I'm having the surgery. I just want you to know, I'm keeping my promise.*

That's the thing about BRCA . . . it'll never go away. Even after I have both surgeries, I'll still have BRCA, and I'll still worry about having passed it on, and I'll always be an advocate for my daughters no matter what their story holds: fighting for getting the information out there, for better research, for better options. We've come a long way, but I really wish, after finding out about this eleven years ago now, I wish I could say that, yeah, there's a blood test, a good proteomics test for early detection of ovarian cancer. A lot of the other things I hoped for haven't panned out yet. But I'll never stop being an advocate for this, hoping that the next generation will have better options . . . The ones we have now shouldn't be the answer, there's got to be a better one.

AMY: One of the things I admire in your film is your patience—with other people, as you talk with them, but also with yourself. Granted you were in your late twenties when you were filming, but even so, there were such real pressures facing you, in terms of contemplating surgeries and thinking about having children. I relate to this in particular because I remember, for myself, feeling an almost unbearable sense of pressure in terms of time when it came to having children. I felt (in our family, with our history) that I needed to move really fast to have my ovaries removed, that there wasn't time. You say now that in your late thirties, as you feel there isn't time left, that you're facing the surgeries now and it's frightening, and sad, but that you also feel a sense of acceptance about it.

What would you say to other women (and I'm thinking about my own daughters in the coming years) who are in their twenties and thirties, struggling with questions about time and pressure and what to do? What was most helpful for you as you waited? What gave you the patience to wait until now for these surgeries?

JOANNA: Many factors contributed to my decision to wait to have surgery, including a network of support that allowed me to shut out the "noise"—internal and external—that was pushing me toward the decision. It wasn't always a cake walk: I remember a moment at one of the screenings of *In the Family* when a woman came up to me, put her hand on my pelvis (lovingly, if a little intrusively) to indicate my ovaries, and told me I was a fool to wait to have them removed. Certainly, it was a condemnation, but I knew it came from a place of deep concern and suffering. How could I wait so long, given everything I know about the devastation of the diseases that BRCA brings on families, including my own? I had an overwhelming desire to have my own children that was as powerful in its fervor and control over my psyche as the nightmares I had about leaving my ovaries intact. All of this was, I believed, within reason, and it helped that it was supported by my incredible physician/guru, Dr. Lee Shulman, who allowed me to utter the words "quality of life" even with BRCA tagged to my name. We agreed that age 40 was an appropriate target for the surgery (my mom was diagnosed at 43) and that we would keep checking in and hoping that I would meet a wonderful man (I did) and have children (I did). There was a lot of faith and handholding and some very dark nights where I questioned this path, feeling a sense of disappointment that I couldn't just "pull the trigger" on the darkest days, but mostly just a strong faith that I was watching myself as carefully as I possibly could (never missing an ultrasound or my CA125 and participating in as much research as possible), and that maybe, just maybe, those years on the birth control pill reduced my risk of ovarian cancer and bought me a little more time. And, most importantly, I was being true to a path that felt right for me. It wouldn't be right for everyone, and had my life been different, had my mom not survived, I probably would have had the surgery earlier.

I also kept researching, learning, questioning, all of which helped. I was looking into the option of freezing my eggs (and maybe some ovarian tissue) when I met my husband. Humor also calmed the storm waters at times (sometimes, I had to laugh at how often I talked and thought about my ovaries—it's just not normal, I once I told a friend . . . I need a new obsession). A supportive and loving family and partner also helped.

Since becoming a mother, everything has changed. The decision is no longer as fraught. There is only one path and an incredible sense of peace and a huge sigh of relief that these surgeries are in sight.

AMY: Before we close, I wanted to ask about this new project you're working on with Rick [Guidotti]. As you say, we've come far, but there remain so many questions. People sometimes wonder, in the wider culture, how much weight to give genetic mutations. There are people who feel this information is life-determining, and others who feel—and I've had this said to me—that there's a lot of "geno-hype" out there, that we make too much of this, overblow it, so I wonder . . . leading into the work with Rick on this new project—how important this mutation has been for you, in defining who you are? And looking forward, how much do you think it will impact your daughters? (As you know, I also have two daughters, who are now 18 and almost 21.) My daughters and I are having this conversation now, but we've been having it for almost twenty years, from how do you talk to them about surgery to all sorts of other things. For you, how big has this mutation been in terms of your identity?

JOANNA: It's an enormous part of my identity. And I don't know if that's going to change with the surgeries, because then the surgeries will also be part of me. The way I've looked at myself, I think that was one of the hardest things. I remember being deeply impacted by Susan Sontag's *Illness as Metaphor* when I started making the film and thinking about this strange way of living both in the healthy world and the illness world ("the night-side of life") with this mutation . . . even though you're healthy, you're still in this pre-cancer world . . . Going back to people saying that this gets overblown—physicians sometimes say, "You're totally overreacting, what are you doing, removing body parts . . ." Recently there's been this whole [debate] accusing women with this mutation . . . of *overutilizing* the health care system, saying, "We're all pre-this or pre-that . . ." I find that very offensive, because there are very few examples out there that are like BRCA, that are so clear cut. BRCA is so astronomical . . . You can't look at these odds, with women having an up to 60 percent chance of getting ovarian cancer, compared to 1.5 percent chance in the general population, and say there's any hype in that. To say that isn't going to totally redefine someone's life, that there's this profound family history and a known underlying genetic trigger, that there's something they can do about it, and they really need to do something, or they could die . . . Considering that hype or overblown . . . that's offensive to me in a lot of ways.

The new project with Rick began when I came across a photo exhibition of his work at the Genetic Alliance, where I was doing a screening of *In the Family*. I was so struck by Rick's photographs . . . I remember thinking, I have this genetic mutation that nobody knows about. I'm not treated differently, unless I reveal it to people. And I wondered what it would be like to have one of the common genetic mutations that are highly visible, where everyone who sees you immediately judges you, whether they mean to or not. It's part of our nature to do that, people who look away, or don't know what to do . . . How do you de-stigmatize that, how can it become part of the common parlance to say, "Yeah! I have a genetic mutation, and I'm taking care of myself"? I think Rick wants to reach out to people who want to look away, to remind them that beauty has a wide spectrum. He wants to tell them, "I just want to steady your gaze." For me, this is a whole other level. I want to learn what it's like, wearing your mutation on the outside as well as the inside. To recognize that we *all* have genetic mutations, and that the ones we explore in the film are just the ones that we know about.

I do think, even though we're not there yet, that the subject of genetic mutations is going to become a much bigger part of the conversation in the coming years, and it won't seem like something we're over-hyping, but instead, preventive medicine will be respected, and funded, and that's how we'll become a healthier society.

AMY: You talk about "steadying the gaze," and I wonder if that's part of the importance of the visual, and what you've done in the film and what you're doing now with Rick, especially when you talk about how what you have is invisible. I feel that myself. I've often thought about the weirdness of "passing" . . . For me it comes at moments like at the gym, do I go into the private little dressing room to change? How much do I expose? Because in clothes, I look like everyone else.

JOANNA: Right!—what do people notice, how much of that is within you, and how much is coming from other people? It's very odd that you can reconstruct breasts and nobody knows whether those are the breasts you were born with or not. It's a strange thing about secrecy, what's visible and what's undetectable.

AMY: So maybe film—particularly documentary film—and the decisions you made about self-exposure—help to make visible things that are often kept invisible, or that we want to keep invisible.

JOANNA: The irony is that I was the classic person coming from the classic family of people who want to keep this invisible. There was a moment when my mom told me that she'd called a friend and told her I had the mutation, and I was upset and told her, "I'm not sure I'm ready for anyone to know yet," and she said, "You just told me that you're making a film about this!" I so badly wanted to keep all of this invisible, which is probably partly why I'm so attracted by Rick's work; he's the opposite, he operates under the mantra of: put it out there, and celebrate, and be yourself.

It was really important for me in the film to show a reconstructed breast, and I was afraid when the film aired [it was on the PBS *POV* series, which is a coveted place for a documentary to air] those segments might get censored, and I was adamant that it not be and, if it came to that, I was fully prepared to go out and wage a public campaign to reverse the decision, and my team and the team at *POV* fully supported me, because I couldn't bear the thought of anyone saying it was prurient to show a medical reconstruction of a breast . . . You can't talk about this without showing it—you have to give people permission to look. I think in both projects, that's really what I'm doing. Giving people permission to ask questions. Giving them permission to look.

*Not long after Joanna and I talked, she gave birth to her second daughter. Several months later, I sent her an email asking a question that was on my mind as I was reading over the notes from our interview.*

AMY: When you and I talked, we noted the irony—in your family and in mine—of all these girls. I know for my part, this is a whole new aspect of the BRCA mutation—the sense of having (potentially) passed it on. I know you talked about this as one aspect of how, for those of us with this mutation, you keep "having" it, even after you've had one surgery or both, or pursued other kinds of prophylactic therapies. Your girls are almost twenty years younger than mine are, but I think we share this acute sense of responsibility, of wanting to make things better for them. What do you project, as you think forward for your daughters? What's your biggest hope, and your biggest worry?

JOANNA: This is the most painful question to answer. I have often said it's ironic that I would have two girls, and cannot help myself from playing out the conversations we will have about BRCA (even though they are far away) and the testing . . . will they both have it? Just one—who? Neither? This is rawer than any emotion I have ever felt about my own BRCA status and echoes the pain I saw so often in the eyes of mothers (including my own) over that "guilt" of passing it on, no matter how unfounded it was.

First, hope: I hope that there is a true preventive therapeutic (like a PARP [poly(ADP-ribose) polymerase] inhibitor) that, if either of my daughters is BRCA-positive, they can take every day that helps out their impaired DNA-repair system and prevents cancer from developing or targets any earlier, unchecked precancerous activity. Will that be a reality? As much as I want to shout *yes*, it may not even hit in their lifetimes; but in my granddaughters' lifetimes, I have great faith that we will be creating preventive, personalized therapeutics. I also hope that discussing mutations will be less stigmatized in the future and as normalized as possible, like discussing your blood type. I hope that my daughters never feel ashamed or different if they have BRCA, or worry that a partner will reject them. If that miracle therapeutic is not here yet, I wish them many in-between options that stave off the clock and don't interfere with their path of self-discovery. One option I am keenly interested in is the removal of fallopian tubes while keeping the ovaries, as a possible interim solution while waiting to have children (using in vitro fertilization). I hope options like these will be covered by insurance and available for every young woman, and not just for those with means.

Second, *worry* . . . [chokes up]. That the information will change them, strip their innocence, cloud their dreams (isn't this every mother's fear?). That they will walk with a sense of doom and that any of their optimism and happiness will be carved away by the worry of illness. That they will feel external pressure to make a decision that does not come from their hearts; that they will suffer, that they will . . . I can't even go there . . .

I need to stop there. I have a baby that's waking up and needs her mama, and she's beautiful and I'm beyond grateful.

One day at a time.

# NINE

# "Why Would You Be Wantin' to Know?"

*Not Talking about Schizophrenia in Ireland*

Patrick Tracey

·······························································································

The Irish have a saying that sums up my frustration with them: When you come to a wall that's too high to climb, toss your hat over the wall and then go get your hat.

But what if the wall is Ireland itself? What if the wall is the cast-iron refusal of the Irish to discuss the mental illness that runs in your Irish bloodstream?

As an Irish American brother of two sisters suffering with schizophrenia—and as a nephew and a grandson as well—I went back there to the source, looking for answers. I met with a few solemn professionals about the discovery of the first gene link for schizophrenia, but when I tried to engage the ordinary Irish on it, all I got was the classic Irish freeze-out from behind clenched teeth.

"And why would you be wantin' to know?" was a typical response.

To say that no one in Ireland wishes to touch the topic is to exaggerate very little. Even though, for 140 years, they led the world with the highest rates of first admissions to mental asylums, you'd have an easier time corralling a bull than getting folks to talk about it.[1]

My own family lore had it going back from Boston to County Roscommon, a landlocked county that's rural even by Irish standards. The county hit hardest by the famine, it's also home to the National Famine Museum. Yes, they have one. And yes, there is a connection with schizophrenia. Famine conditions can almost double the chances of onset, according to the Dutch Winter Hunger Studies.[2]

For the best of reasons, I needed to talk about it. Mostly I needed to talk with others about how there's nothing quite like the heartache of the loss that comes with this psychosis that dawns rapidly in late adolescence to early adulthood. In our family, it appears at an age when most people have barely begun to live their lives. Two of my four siblings—gorgeous girls in the bloom of life—went rapidly mad in their early twenties. Thirty years later, they've never recovered.

What led me to Ireland in the first place was a modest finding in the field of molecular genetics, the discovery of an abnormality in a gene coding for a protein called dysbindin that's involved in the synapses, the points where one neuron connects itself to another. Normally, the discovery of a chromosomal abnormality would be no cause for excitement in our lives, but the marker was found in blood samples taken from a few hundred families with schizophrenia in County Roscommon in the 1980s.[3]

From what my mom said, the disorder in our own family was directly traceable to her grandmother's grandmother, a young woman who was coughed out of the small parish of Cam-Kiltoom in southern Roscommon, not far from Athlone, in 1847, in the middle of the Great Irish Famine.

Her name was Mary Egan, and when she landed in the New World 165 years ago, having escaped the famine in a so-called coffin ship, the *Anglo American*, she got out with her life, if not her mind. Mary Egan was already away with the fairies, as the Irish say, meaning she was madder than a waltzing mouse, when she hit the wet docks of Boston. Whether her mind collapsed on the high seas or back home in famine-ravaged Roscommon remains a mystery to this day. What is beyond doubt is that she unpacked her errant gene pool in Boston and passed it down through the generations of New Englanders, until it pooled in our immediate family.

Mom was the keeper of this history that came down from her own mother's side—and mostly kept it hidden from our tender ears. When I asked her dad, my grandfather, about his own mad wife, my grandmother, May Sweeney, who I never met, all he ever said was that her side of the family was "away with the fairies" and that the women were the carriers.

Hoping to dodge it, our mother set her sights on a career as an attorney. Long before women were being admitted to the bar, she was trying to sidestep the traditional routes of motherhood in the 1950s. Mom wanted to have children, but she feared she might pass on whatever it was that had sent her mother and brother away with the fairies in their twenties.

Mom's plans to remain childless came a cropper when she caught the eye of the wrong man. And not just any wrong man. Dad was just the right wrong man to trigger her errant gene bank. An old adage is that "nature loads the gun and nurture pulls the trigger." In hindsight it's easy to see how Dad pulled Mom's trigger and shot off two of five of us.

As he used to tell it, Mom was a black-haired, blue-eyed Irish American beauty causing a major traffic jam in Providence when he happened along in the 1940s. A tall, handsome young stranger, he told her to scoot. Mom protested, he insisted, taking the wheel as she slid into the passenger's seat. That was it. Dad had that dash, and it could be said that he literally drove her wild.

Mom fell head over heels, and within three years, he had her married and pregnant. Big mistake, because through her womb would come this very odd, late-blooming psychosis that cascades through our bloodstream like water through a swollen river. Her brother had rapidly developed the psychosis in his twenties, as had her mother, whose own mind had left her as swiftly as a dark shawl that slipped from her shoulders on a leafy day in the autumn of 1929.

The story of how May Sweeney went for a walk one day in her Sunday best was rarely shared in our family, no doubt because one telling was enough. She wore a black chiffon dress, cloche hat, and pearl necklace. By nightfall, her husband worried that she was nowhere to be found.

That was our grandfather, Big Jack, who was pacing the floors of his Hannah Street home, waiting for his wife. By mid-evening her dark figure had appeared, coming up the walkway, but something was off. May's hat was cocked sideways and she was barefoot and clutching her shoes instead of wearing them, her small feet reddened sore from hours of walking.

Jack said nothing, but dropped his teacup as she finally smiled: every one of May's teeth—brace for it—had been pulled, her mouth gorged with blood. Later it was learned that she'd actually paid an unscrupulous dentist for the surgery. She thought it would silence the "voices" in her head.

She spent the rest of her days in the company of her bedeviling voices in a state mental institution in Rhode Island. We never met. She died three years before I was born, but a few years later I did get to meet her son, my Uncle Robbie. He had it too.

Robbie was carried out of the same Providence home on a stretcher two decades later, having come home from the war hearing voices. I was with my mother and sister, visiting him on his own state farm one day. I was about 5 and Austine was about 6, and he took the two of us out into the field, presumably to show us how to milk a cow. Instead he did something unexpected out there with the cow. He slapped me hard across the cheek for no apparent reason.

When we marched in from the field, I showed Mom my red cheek, and we were whisked away. It was on the way home that I first heard my mother use the *s* word. "Your uncle is not well," she said, tersely, from the front seat of her big Cadillac, puffing on her long, slow-burning cigarette. "He has the schizophrenia."

She said it all went back to Ireland, and there was a cure, water from a healing well over there, and that we'd go one day to fetch this water. That day never arrived. Robbie was gone within two years, dead from a heart attack, and the family was without schizophrenia for a full decade.

I forgot about his odd behavior, mainly because my mother put the kibosh on any talk of it, right up until the mid-1970s. That's when the madness came roaring back, coming first for Michelle. Our artsiest older sister, Chelle, as we call her, was a teenage Polaroid model who'd studied at the Sorbonne in Paris and was now on the stage in New York, in her mid-twenties, a blithe spirit and budding actress, working Off Broadway.

Chelle has hopped a bus home from Manhattan for the weekend, and I am sent to meet her at the Providence bus station. I am 17, slightly stoned, and late picking her up as I was told to do. By the time I get there, I see my sister singing and dancing around the bus station like a demented commuter. She's leaping and twirling around a news stand, disappearing and then reappearing, her darting eyes challenging any man to dance with her.

Michelle takes my arm, and on the ride home in my mother's Ford Pinto, she explains that she has just broken up with the film star Warren

Beatty. She says she left him for Jesus Christ, whom she believes to be the Romanian tennis star Ilie Năstase. She blurts all this out with a voice so full of the God's honest truth that, for a fleeting moment, I wonder if this isn't all an act, like a new role she's rehearsing for some stage performance back in the Big Apple.

She says the whole world will be at her wedding day ceremony and that her four sibs—Elaine, Seanna, Austine, and myself—will have seats of honor at the head table with Mom. Dad, separated from Mom for several years now, isn't even invited. He'll drink too much and spoil the day. Then she tells me I'm the best man—evidently, because I'm the Biblical John the Baptist. All righty then.

I remember having an ear infection that made me delirious as a child, but only temporarily. I assume Michelle is suffering from something similar, that she'll see a doctor and right herself in a few days. Then Mom comes home and, as her eyes fall on Chelle dancing around our home, she uses that word that I haven't heard since Uncle Robbie died. "It's the schizophrenia," she says.

"Huh?"

"The schizophrenia."

To me, the word sounded like a rake that scrapes over you and leaves behind scratches. I never imagined that thirty years later, Chelle would still not be right. She lives in a group home and has never recovered from her dual diagnosis: manic depression and paranoid schizophrenia.

And two years later, a second sister would follow Michelle. This time it was Austine, one year my senior, a sweet, freckle-nosed sister with all the personality and popularity a kid could have in the 1970s. Austine had taken a year out of college to work at Disney. I happen to be there again, this time at the airport with my mother, when she comes back. As she steps off her flight, shoulders hunched, body rigid, head bent forward, we can see that the old Austine has not landed. In her place is this new person, this spaced-out zombie.

Austine isn't singing and dancing like Michelle was at the bus station, but she's much further gone, lost in some deep oblivion. Her eyes are leaden and lifeless, her body stiff as she hugs us. Her face is without expression, her whole affect flat, her once bright and shining being vaporized now. Austine cannot answer a single question with any intelligence. She answers "yes" or "no" only, and gets it wrong half the time.

"Austine, are you home?"

"No."

"Austine are you okay?"

"Yes."

She isn't mute, just struggling with basic mental cognition. At the same time, she seems frightened in her aphasia, preoccupied by some presence beyond our ken. Like Chelle, she has never recovered. Like Chelle, she spends her days shoegazing in a group home for schizophrenics in Boston.

Mom, at the time, couldn't face it, insisting her baby girl was "just going through a rough patch. She'll pull out of it."

I saw the denial for what it was, a silent scream from the bosom of a mother's worst fears realized. When one daughter dropped, it was hard enough. When the baby girl followed, Austine's mind collapsing like a second Twin Tower, Mom could not deal.

Weeks later, the two are home alone together, having Swanson TV dinners in Mom's bayside condo. I can see Mom sitting there and trying in vain to jump-start the conversation with her dumbstruck daughter. Impossible. Austine just looks at you like a cow, her eyes lifeless. She chainsmokes but is unable to respond to simple questions. It's exhausting. Whatever you say, Austine won't say boo back, nothing beyond a simple "yes" or "no." You feel drained by the effort.

I can imagine Mom peering across the flowered centerpiece on the table between them in her condo. There it was, the loss of what she loved most deeply, her baby girl, a second girl gone like all the others. I can see Mom moving from memory to memory, the silence between them broken only by the cry of seagulls over the Narragansett Bay. At some point she must've felt the worst headache of her life. A spurt of blood had punctured an artery, flooding her brain. Mom heaved her last sigh and died where she dropped, more or less at Austine's feet.

In its own weird way, schizophrenia got her in the end, too, but it did not end there. The wake at Skeffington's funeral parlor features Mom in her coffin to the left of the five of us, taking our place in the receiving line. Elaine's 30; Michelle and Seanna, fraternal twins, are 28; Austine is 21; I am 20; and Mom didn't see 60.

I never bother to place the bet my mother lost. Only Seanna rolls once and wins, with Chris, her only son, who's also been like a son to me.

For me, the chance that I might be holding the gene code that hovers like a Predator drone over future generations has never been worth the risk. The threat of madness would always be there, like a pillow that'll be used to snuff a child out in late adolescence.

I dodged fatherhood and drank a lot instead. I tossed it back because booze made me forget about things in a way no therapist could. I kept my legs spinning on bar stools until 2001. Then one day my own meds turned on me.

It was in my flat above a fish 'n' chips shop in central London. I'd just emptied the last of my whiskey to the dregs when I began to hear intermittent clanging sounds. While I wasn't hearing voices—no direct or indirect speech from disembodied personages, as is the case with more robust auditory hallucinations—these vague sounds appeared to be imaginary. Hearing things that can't be heard by others could only mean one of two things: end-stage alcoholism or early-stage psychosis. This was a club I could not join, so I ended a thirty-year run and climbed into the rooms of AA. Who knew there were six hundred meetings a week in London?

Over the coming days I was welcomed into the homes of the loose-knit community of recovering English drunks, and never looked back. I remember feeling better than ever—though still pretty blinkered—as I celebrated my first year sober. It was June 21, I had my one-year chip, and I joked ironically with my sponsor that I had to pick the longest day of the year to quit.

"Ah, but the shortest night, mate," Cockney Mick said, clinking teacups over a full English breakfast in a worker's café near Highbury Fields.

I was sober and, beyond that, with new mates like Mick befriending me left and right in London's AA, I was beginning to grasp the grand paradox of recovery—that you're grateful to have a problem you can't solve alone. I isolated with alcohol for so long that it was just great to reconnect with the human race, but I knew—and Mick knew too—that I wouldn't stay sober long unless I dealt with the loss.

Within three weeks, Ireland called. The "call" came in the form of the July 4, 2002, announcement in the *New York Times* that the ever elusive gene link for schizophrenia, always suspected but never found, had been located in blood samples drawn from people with schizophrenia in

Ireland. Not just in Ireland, but in blood drawn from 258 people in County Roscommon.

The news was eerie and thrilling at the same time. Finally, after years of desultory efforts to crack the genetic code, here was the first legitimately good piece of news. For all that time I'd seen my family's life as a rough draft of a story that had gone nowhere. Now the place that had cursed us was coughing up answers. Something worth looking into had poked itself out of the Irish muck, just as I was in neighboring England.

The research was solid gold, peer-reviewed, and published in the *American Journal of Human Genetics*, and now reported in the *New York Times*. Soon it was replicated in other studies and would quickly be followed by the discovery of more susceptibility genes outside Ireland.

I knew nothing about genetics but was riding the foam of hope that one day there'd be a medical answer. Irishly superstitious, part of me saw it as a sign, news that might one day help unlock my two sisters from their psychoses. Ultimately, I was naive, but it's never too early to hope when hope is all you've got.

From my flat in central London, I felt a strong gravitational pull to cross the Irish Sea to see what made my family's blood so spongy with this schizophrenia. I was intrigued enough to want to go back 160 years to the land of my ancestors, and isn't that the power of the genome? It allows us to go back in time to try, at least, to solve the problem.

The more information I was able to gather ahead of time online, the more curious I became, with certain factoids like the strong associations between schizophrenia, on the one hand, and famine, emigration, substance abuse, and older fathers, on the other—factors that were prominent in the peasant Irish experience through a famine economy that stretched over the centuries.

And then I became immediately fascinated by a forgotten piece of Irish history—a true medical epidemic of what was called lunacy in the parlance of the day. That lunacy—largely schizophrenia and bipolar illness—traumatized Ireland in the nineteenth century and half of the twentieth century and rolled across the Atlantic like a great wave of madness.

My family, as it turns out, was not atypical. In the nineteenth and twentieth centuries, the Irish were filling up the asylums on both sides of the sea—in Dublin and London and Boston and New York, the asylums were packed with Irish. (Today rates have leveled off, so that if you're Irish, you're no more likely than anyone else to develop schizophrenia. At one time, however, you had a real head start.)

By the summer of 2006, I was all in. I bought a camper van and lived in Ireland for four months. First I met with Dr. Dermot Walsh down along the Grand Canal at the Health Research Board. It was Walsh who initially had the idea of collecting blood samples in Roscommon. I remember feeling a stir of expectation about what Ireland's grand old man of schizophrenia research might say.

In the conference room of his office suite, he told me about the work he'd done with Dr. Kenneth Kendler, an American, and how it had eventually led to the discovery of this gene marker. I was hoping he might puff his chest out with pride at the importance of his findings, but the twinkly-eyed epidemiologist folded his hands and spoke in soft, measured tones about the modesty of his findings.

"It's quite clear that its effect, like some other genes that have been since discovered, is quite small and you will only get this effect in a small proportion of individuals," he told me. "How it works and how it operates is another day's work."

Given the depths of the medical mystery, this news was a bit of a letdown. For some time, science had been touting the decoding of the human genome. I myself had been drawn in by a great deal of excitement in the 1990s about what molecular genetics might be able to tell us about hereditary or genetic contributions, not just to schizophrenia, but to all major mental illnesses.

Walsh explained that giant strides were being made in research into rather simple diseases (genetically speaking) such as Huntington's, cystic fibrosis, and so on, but venturing into the dark corners of DNA for schizophrenia would take more time. He reminded me that it was known at the outset that in this case we were not dealing with something genetically simple in the old Mendelian sense. Still, an aura of expectation had developed around the techniques of genetic analysis, if only because early studies had confirmed the impression, going back to the birth of modern psychiatry, that schizophrenia ran in families. Later adoption and twin studies have strongly indicated that genetic factors make an equal contribution with upbringing. Finally, the Irish Study of High-Density Schizophrenia Families, reported in July 2002, produced the first evidence of a gene that's expressed in neurons in many areas of the mouse and human brains. Poetically, it was in the original Roscommon blood samples that this discovery in the nationwide study was made.

When I asked Dr. Walsh about these families, I learned that personal contact had been made by calling unannounced at homes. Face-to-face interviews were then conducted with those who suffered and their first-degree relatives. Researchers talked to parents, siblings, and children 16 or older.

Even though these were blind studies—and I could never hope to speak with the participants—just learning about these in-person interviews unleashed in me an indelible desire to know and meet other families whose experience was as pronounced as our own. I was a stranger, though, and I was not offering the happy, chirpy conversation the Irish have come to expect from Yanks over on golf and pub tour holidays.

I wasn't that guy. I knew that going in. The Irish can be an insular people. What I underestimated was the borderline hostility that would meet a man on my mission. I've only read about families as mad as my own—I've never actually met one, though I've heard distant echoes of other avalanches—so I felt I had a right to be asking the indelicate questions. But just to utter the word *schizophrenia* on the back-country lanes—the *boreen*—is to hurl the cat among pigeons: people scatter.

To be fair, the *s* word splits the air with fear almost everywhere. It's just that the Irish seemed more reflexively standoffish. Closed-off body language hints at hidden fears; the back of the hand, the uncertain invitation to clear off.

Clearly I was touching nerves as raw as a bad tooth—nerves that go back centuries. In old Ireland, a family history of mental illness could seriously crimp your marriage prospects, grinding you into deeper poverty still. But even in modern, buzzy Dublin, the world's most loquacious people fairly groan at the mention of mental illness. It wasn't a clash of opposites. Nobody pretended *not* to know what I was talking about when I mentioned the famine-driven epidemic of severe psychosis that had fallen (conveniently?) into the forgotten pages of Irish history, a fascinating story in its own right. It was just that, suddenly, they all had to be elsewhere.

Admittedly, I was on the darkest of tours, jerking blindly along in my second-hand Nissan Vanette, my four-month home on wheels, casting about for crumbs of clues from 165 years ago. I was able to scare up a ten-minute cup of tea with one mother whose mad son had died a few years before in Cam-Kiltoom, the small Roscommon parish of my ancestors, but she was the exception that proves the rule.

I made an appointment to see the president and executive director of a prominent group that deals with schizophrenia in Ireland—Schizophrenia Ireland, a leading nonprofit group that advocates for greater awareness. I explained that I was there doing my own personal research, that I was hoping to hang out with my counterparts, the brothers and sisters and mothers and fathers and daughters and sons whose hearts crackle with the same ache.

More silence. More concerned glances. "It just wouldn't be appropriate," the executive director said at length. She said I was welcome to attend their annual conference, but I was not welcome to rub skin with the others at any get-togethers of the relatives group. I was not Irish, so they weren't taking chances.

I sat there in silence, an island of resentment all to myself. Composing myself outwardly, I nodded mutely, not wanting to touch another nerve. I felt embarrassed for misfiring. I thanked them, wished them luck, and ducked down and out along the canal.

Back in my van, I motored around the island nation, circling around in a widening gyre on my way out to find that healing well my mother had told me about in my boyhood. I found the well water, but mostly I just saw more ghosts from the window of the '94 Nissan Vanette I'd purchased back in London. The second-hand van did the trick for the wettest summer on record in a nation known for rain. For months, it fell like stair rods on the tin roof. And each day I got out long enough to stare into the clenched jaws of another clammed up Irishman.

No wonder Sigmund Freud observed that the Irish were the only people in the world who would not respond to psychotherapy. They'd hardly even broach it. In almost every encounter, when I mentioned my quest, they looked at me as if they wanted to run me off like a would-be poacher.

A stubborn, defiant streak in my own nature kept me going until mid-November, when it became cold enough to hang meat in that van, and still I got nowhere. The ten-minute cup of tea with one possible distant relative whose son had a diagnosis meant everything—but, again, it was the exception that proves the rule.

After this, I wrote an overly dry book explaining how famine-era maternal malnourishment nearly doubled schizophrenia rates. It discussed famine-era late age of paternity driving rates higher still, in all likelihood. But the one thing I wanted to explore most of all eluded me: the anguish that tears us apart. I never got to have the handshaking experience with other families that our mutual problem would warrant.

Alas, life, as the Irish would tell you, is a dance full of unexpected turns. Somehow my little tale of woe about getting snubbed in the grass-green land of my ancestors found its way into the hands of a small but devoted readership with families like mine back home. These Americans differed from the Irish in one key respect: they were eager to talk. A flood of emails, some from families that rivaled my own, told me so.

I realized something I'd never considered—that while nothing in old Ireland may be less favored than self-revelation, Americans are modern enough to embrace it. One email that stands out for me was from an Irish American woman who wrote to say how she and her very best friend, another Irish Catholic woman living in the Boston area, had met every day for thirty years for tea. The two women, being the best of friends, rarely kept a secret. As tight as they were, they talked long like the Irish over everything under the sun, except the one thing under the sun that can't seem to be talked about.

"How's Johnny?" her friend would ask.

"Oh Johnny, he's fine. And how's Bobby?"

"Oh Bobby? Oh, he's fine too. Grand, Bobby is."

And there it was left each day, a clipped exchange, until her best friend died and she went to the wake, and there in line in the funeral home she saw that it was not all grand with Bobby. He was talking to his invisible head friends while his mother was laid out for viewing.

She realized that they both had boys at home who heard voices, but neither mother could talk about it to her closest friend. For thirty years they never spoke of it—never once—so heavy was the lump of shame.

"I am writing to tell you," the one surviving woman said to me, "because then and there I vowed never to remain silent."

I treasured receiving emails like this. Then one day, a few months ago, a letter arrived from my publisher saying my book was going out of print. Soon afterward, the emails began to dry up, too. I felt the old void creeping back. It weighed on me that I'd never met the other families. My sisters were squared away in group homes, so I set my sights on an imaginary family that I might meet in a thousand days ahead, a thousand miles away, in the cold clear light of day. Who knows? Their woes might be worse than mine. I might be useful to them, too.

So I'm pointing my vehicle across America this time, looking for families that mirror my own, the most multiply afflicted, that agree to meet ahead of time. I'm using my first effort as my calling card for the next, with the conviction that, if there is not an excess of caution, then what we discuss in coming together might do far more than what we can never seem to say in silence apart.

I'm feeling better already, a bit less restless, and a lot less resentful that the Irish dogs were barking as my caravan moved on. I'm over it. I'm flinging my hat over that tall wall of silence that divides us. And this time, I'm crossing my own country to get it.

# TEN

# Help Wanted
Michael Downing

..........................................................................................................................

Seeking reliable storyteller to write the next chapter of the story of my life. Successful candidate will be familiar with the relevant history detailed below. Brevity, wit, and originality valued; medical expertise and experience—not so much, though professional credentials will not be used to disqualify candidates.

All responses deemed leastways worthwhile will be used at my discretion to inform the impending choice of continuing or abandoning the prophylactic therapy begun in 2005 to prevent my sudden death. Please submit advice, recommendations, and any supporting data, qualifications, or references to michaeldowningbooks.com.

I hereby absolve all respondents of legal liability regarding my health and longevity. I reserve the right to characterize, quantify, and reproduce all or part of each response for as long as I shall live.

---

On a warm night in July 1961, my father got out of bed. He walked into the bathroom adjoining the bedroom he and my mother shared and coughed a couple of times. My mother found him sprawled out on the linoleum floor. She woke a few of my elder siblings, and an ambulance was called, but she'd known as soon as she saw my father's prone body that he was dead.

When the first-responders arrived, my mother was in the bathroom, between the bathtub and my father's body. She wouldn't move until his body was carried away. In many ways, the family history I inherited never moved beyond that moment. The specter of sudden death was always in front of us.

My father was 44 when he died. He had no personal history of life-threatening illness. His three elder siblings remained alive for decades. Without benefit of an autopsy, the medical examiner attributed his death to a massive heart attack. The bigness lent a heroic aspect to an event that hardly needed inflation. My father was a beloved public figure in Berkshire County and throughout Massachusetts, president of the Association of Business and Commerce, a successful mediator and nurturer of the region's intersecting industrial, political, and labor interests.

I was the youngest of nine children, and my elder siblings were the sort of kids who regularly turned up in local newspaper stories about high achievers. My father and mother were famously devout Catholics. We recited the Rosary every night on our knees, often in the company of an unsuspecting friend or first date who'd accepted an invitation to a free meal.

In family conversation and public tributes, the massive heart attack was construed in religious terms. Diagnosis: God's will. But no one wanted history to repeat itself.

The family pediatrician started listening more closely to everyone's heartbeat. There were rumors of heart murmurs, and dietary changes specifically targeted at the boys. Maybe heart attacks really were understood to be singularly threatening to males, or maybe the life span of women was, like their opportunities to play a sport or earn a decent wage, not a priority. My mother religiously prepared three meals a day, but she didn't tolerate discriminating palates, so for several years, all of us—girls and boys alike—drank no-fat powdered milk, ate oatmeal at least three mornings every week, and, hardest of all for me, lost our butter privileges.

The ban on normal dairy products eased when Gerard went away to college. He was sixth in the sibling lineup, and though six inches shorter, by every other measure he was a ringer for my father—appearance, disposition, verbal dexterity, and general verbosity. Before we became acquainted with our DNA, we practiced a kind of folk genetics based on observable traits and tendencies. Although something of both of my parents had been detected in most of my siblings, a few of us were less evidently hybridized. My sister Peg (number 5) was second only to Gerard in resembling my father, and Mary Ann (number 4) and I were perceived as almost exclusive inheritors of the qualities (admired and not) of my mother.

My mother wanted all of her children to thrive, but Gerard was the apple of her eye. His habit of enjoying an apple with thick slices of cheese and mayonnaise must have recalled my father's incautious eating habits. I am sure it informed her unscientific prognosis: He was most like his father, most at risk.

On a snowy morning in December 2003, my brother Gerard woke early and went outside to clear a path to his morning newspaper. Like my father, Gerard was not a huge fan of physical labor. Instead of a shovel, he went outside with a broom. The youngest of his four children found his body in the snow later that morning.

Gerard was 53 when he died, the longtime district attorney of Berkshire County, committed Catholic family man, admired and beloved and too young. He was overweight, and his cardiovascular fitness had been a cause for concern. No autopsy was performed. The medical examiner considered his death a clear and simple case: massive heart attack.

In the family history, Gerard's sudden death was the fulfillment of a tragic legacy, the passing of a too-fast burning torch from father to son. But about a decade before Gerard died, his story had been taken up by an unrelated team of storytellers, geneticists at Harvard in the early stages of decoding heritable protein mutations in human DNA. They were writing a secular salvation story, a counter-narrative to the tragic stories that end when a heart suddenly stops beating.

I don't countenance the existence of god, but those who do assure me he works in strange ways. So do geneticists.

Gerard's third child, Nate, early in his life, experienced episodes of profound dehydration and dizziness, which landed him in the care of a pediatric cardiologist in Boston and a diagnosis of hypertrophic cardiomyopathy, a progressive disease whose most famous and often first symptom is sudden death. HCM is often diagnosed, postmortem, in young athletes.

HCM is, colloquially, the coarsening of the muscle fibers in the heart. This can be observed during an echocardiogram as a thickening of the wall of the heart, enlargement or distortion of the ventricles, and disarray in the sarcomeres, or muscle fibers. The diseased heart is less efficient, compromising the operation of the aortic and mitral valves.

Stiffened ventricle walls prevent normal pumping of blood, especially during exercise or exertion, and complicate the heart's capacity to regulate the heartbeat. The stress on the increasingly inflexible heart eventually causes the heart to dilate, to become larger and sluggish, presenting complications not only for breathing but for the function of all vital organs. Sudden death can result from an instability brought on by unsynchronized operation of the heart muscle—fibrillation; by bouts of irregular and unsteady pacing of heartbeats—arrhythmia; or by wild upswings in the beats per minute—tachycardia.

There is no cure for HCM. Beta-blockers are often prescribed for disease management, despite their well-documented incidental tendency to exacerbate heart failure in patients with compromised myocardial function. There is no evidence that these drugs slow the progress of the disease or mitigate the likelihood of sudden death. They were of no evident use to my brother Gerard. HCM, its idiosyncratic rate of progress, and the relationship between the degree of observable thickening or coarsening of the muscle fiber and the likelihood of sudden death are not confidently understood.

As a teenager, Nate was outfitted with an implantable cardioverter defibrillator, a miniature version of the paddles used to shock technically dead people back to life. An ICD is a stopwatch-sized electronic device installed just above the pectoral muscle and hard-wired to the heart with electronic leads that are threaded through the superior vena cava, the large vein that carries deoxygenated blood from the upper body back to the heart.

After the ICD was installed to prevent Nate's sudden death, the Harvard geneticist, a colleague of Nate's cardiologist, approached Gerard. As a result of that meeting, Gerard asked his siblings to send blood samples to the geneticist's laboratory for analysis. Several of the siblings responded to that Help Wanted ad, asking for nothing in return and, for many years, getting exactly that.

In early 2004, I was 46, 5'10', weighed 155 pounds, and my blood pressure and cholesterol numbers were reliably at the low end of the normal range. By all conventional cardiovascular measures, I was as healthy as I had been since college. This tallied with family history. My mother was almost 80, upright and active, despite the trauma of a few surgeries in recent years. By all appearances, my future was unrelated to the fate of my father, Gerard, and Nate.

Well, by all appearances but one. By the end of March, my sister Mary Ann, the sibling with whom I had the most in common and whose sensibility I trusted most, was sporting an ICD of her own.

In the weeks following Gerard's death, Mary Ann had tracked down the Harvard lab to which, years earlier, she'd submitted a blood sample. The lead geneticist told Mary Ann she'd read of Gerard's death and was eager to meet. The geneticist and her colleagues had identified a mutation in a myosin-binding protein that was pathogenic, autosomal dominant (heritable from a single parent), and clearly identified with HCM. At their first meeting, the geneticist offered to do a DNA test for Mary Ann and referred her to a cardiologist and an electrophysiologist with whom she worked. Based on family history, even before the DNA test, the geneticist urged Mary Ann to consider immediate ICD implant surgery.

After years of silence, the Harvard lab was sounding an alarm—and offering free DNA tests to all of the Downing siblings. Mary Ann, who had three children, had pressed for and successfully won free testing for the sizable next generation, as well. This was more than anyone in my family had been promised when they'd submitted blood samples a decade earlier, but it was less than we were owed.

For at least a few months, and maybe several years, the geneticists at Harvard had known that the mutation they had identified in Gerard and his son was inherited from one parent. They knew Gerard's eight siblings, and their many offspring, were potentially at risk of sudden death.

There is a fine but hard line, a cardiologist associated with the genetics team later told me, between pure research and clinical treatment.

Maybe so, but it was observed purely in the breach—by the Harvard lab, or by my brother Gerard, or maybe both. The geneticist told Mary Ann she'd sought out my brother and told him about the pathological mutation. If so, Gerard never mentioned this to any of his siblings before he died. Maybe he was waiting for an event when many of us would be together, or was trying to imagine how to break this news to his other children, or maybe he didn't believe the mutated gene was definitive. Someone with a more sympathetic imagination than mine can explain the laboratory's choice to inform one carrier of an inherited mutation and none of the eight members of his immediate family, some of whom were almost certainly at risk of sudden death.

My mother was not swayed by the gene-based theory of family history. She was willing to concede that Mary Ann's physician had noticed some thickening of her heart muscle a few months earlier, but my mother believed in the singular father-son lineage she'd seen in my father, Gerard, and Nate. Plus, she had never been a big fan of Harvard, especially after my brother Joe and I went there as undergraduates instead of choosing one of the Catholic colleges our elder siblings attended.

After Mary Ann's implant surgery, my mother said to me, You're not getting one of those things, are you?

I said I wasn't sure I needed one.

After dithering for a couple of months, I had just Fed-Exed some of my DNA to the Harvard lab.

You don't need one, my mother said. You're just like me.

When a nurse called and told me I had tested positive for the mutation and referred me to the doctors who'd outfitted Mary Ann, I still had my doubts. I asked Peter, my partner of almost twenty-five years, the person who loved me best, the man whom I trusted most, what he thought I should do.

He said, What would you want me to do if I were the one with the bum gene?

You are driving a reliable car on a highway at night. The passenger seated beside you is talking on her phone. It is foggy. You are in heavy traffic, driving at the speed limit. You see nothing but the red taillights ahead of you, a steady stream of headlights behind. The passenger suddenly drops her phone, grabs the dashboard with both hands and screams, "Stop the car!"

Nothing.

Your passenger is still screaming. "Stop! Somebody put a bomb in your car!"

That's what it was like to be told that I had inherited a pathological mutation whose first apparent symptom was likely to be my sudden death. It made for riveting conversation, but it did seem far-fetched. I didn't slam on the brakes. Stopping seemed risky, and I had places to go, books to write, courses to teach. Proceeding with caution seemed the right thing to do. Right?

But if your father had died in an exploding car, and your brother had died in an exploding car, you might slow down, pull off the road, and have someone check under the hood.

After the DNA test, doctors treated my family history as urgent and incontrovertible evidence that I was doomed. I preferred the old version, but even after I'd passed her EKG and stress tests, the cardiologist sent me upstairs for an echocardiogram.

The technician wielding the wand that transmitted pictures of my heart to the little TV screen stopped at one point to let me turn and look at my heart. It was the first time I'd seen it. It looked normal, beating away. And then a doctor barged in with two interns and announced that he wanted to have a look. He never looked at me. He looked at the TV, scowled, took a few steps forward and pressed his nose to the screen.

I thought, *The ticking bomb.*

He said, There's nothing there.

The technician was nervously skidding the joystick around my chest, responding to commands barked out by the doctor.

He doesn't have it, the doctor said, disappointed. He led his attendants out of the room.

The cardiologist later assured me I did have it, that there was some thickening evident on the pictures, and some misalignment in the sarcomere. In her notes from this meeting, the cardiologist characterized me as "a 46-year-old gentleman in excellent health, and with no symptomology of active cardiovascular disease." But my apparent health was used against me, to heighten the probability that my first notable symptom would be my sudden death. She sent me down the hall to meet the surgeon.

You can't quit while you're ahead if you are never ahead. And it is hard to get ahead of the prospect of your own sudden death.

The first foray into my chest and vena cava took place in July 2004. A Medtronic-brand ICD was implanted in the natural pocket below my collarbone. The electronic leads were anchored into the wall of my heart. I had a four-inch scar, and a warning about electronic security gates, car batteries, and other devices with magnets powerful enough to interfere with my new accessory.

My only complaints were the prominence of the ICD—it bulged out on my chest, as apparent as a sheriff's badge—and a small cut or pimple I felt on my back after surgery and reported to a couple of nurses, who confirmed the presence of something tiny and black back there but didn't consider it worth discussing.

The random pains I reported during the next few weeks weren't real pain, the surgeon assured me, as I didn't have nerves in those locations. Whenever I reported discomfort, stiffness, or imbalance to anyone at that hospital during the next month, I was told it was normal.

After Labor Day, I taught my first classes of the semester at Tufts, and my ICD started to disappear. I called the surgeon's assistant to report that the site was red and puffy enough to make the device invisible. As I didn't have a fever (I'd checked), she told me she would try to fit me in for an appointment the next day. By then, I could see thick red lines under my skin, radiating from my device down my left arm, and my chest was hot to the touch.

The staph infection I was given with the ICD had been brewing for seven weeks. It found its way into the pocket in my chest, bubbling up like activated yeast into my vena cava and sliding down the hard-wired path to my heart. To prevent my immediate death, the surgeon explained, he had to perform an emergency explant surgery and remove the leads and the device, which he'd implanted to prevent my sudden death.

I woke with a new scar, a hand-operated pump attached by a tube to the empty pocket for draining pus and other infected detritus, and no protection from the possibility of sudden death. The many doctors and nurses who came by pronounced the whole operation a tremendous success. At one point, Peter led my mother into my room for a brief visit. She took one look at me and said, I think this whole gene thing has gone too far. Before I was discharged, a PICC line was threaded from a vein in my arm to the top of my superior vena cava, and for the next six weeks, three times every day, I devoted about forty-five minutes to self-administering three vials of antibiotic chemotherapy.

The surgeon and the first two infectious disease specialists I met doubted that the hospital was the source of the staph infection. My request for a DNA test of the staph, to determine whether the same strain had been found in other patients, was dismissed, as was my initial request to speak to the head of infectious diseases at the hospital. At that time, I was in an examination room with my shirt off, so I decided not to put my shirt on, backing up traffic in the waiting room until I met the head guy.

He was a disappointment. He showed me a computer graph that allegedly proved this hospital's rate of infecting patients was precisely the same as the rate at every hospital in Boston. The word *collusion* came to mind. For starters, we both knew that my hospital-acquired infection had not been recorded as hospital-acquired. He reminded me that, no matter what hospital I was in, every foray into that pocket in my chest doubled the likelihood of infection.

Before the third surgery, in December 2004, I submitted a number of requests for precautionary measures. Each request was met with an impatient nod, but I did get a presurgical prescription for an antibiotic booster, and an early antibiotic drip in pre-op, and several topical treatments not used during my first two surgeries.

I woke up with a new scar, a sore chest, and a notably slimmer, less prominent ICD manufactured by Guidant, which the surgeon told me he'd special-ordered to improve the fit. Even I considered this operation a success. And for more than two years, it was. There were a few unsettling moments, notably an article about ICD recalls, which were not uncommon, in which a Medtronic spokesperson estimated that the risk of removing an implanted ICD was thirty-three times greater than the risk of living with a potentially faulty device. Each time those wires were inserted or removed, there was some insult to the vena cava, which occasioned scar tissue, which reduced the amount of space available for blood flow. Plus, a nick to that vein during surgery could cause undetectable internal bleeding, which had resulted in apparently stable post-op patients bleeding to death.

To test the seven-year battery on my new ICD, twice a year I rehearsed my death. I was given a sedative, the surgeon programmed my device to induce a fatal arrhythmia, and we both hoped the ICD would deliver a life-saving shock. I never felt anything.

But in October 2007, I got a nasty shock of recognition. A *Wall Street Journal* headline announced "Medtronic Pulls Defibrillator Wires

Off Market." The subheading was worse: "After 5 Patients Die." Medtronic initially estimated that 2.3 percent of patients with its Sprint Fidelis leads would experience a problem—unneeded shocks, abrading and punctures caused by fraying wires, or failure to receive a shock at the fatal moment. An independent hospital-based study in Minnesota turned up a failure rate of 70 percent.

Sprint Fidelis rang a bell.

The first time I called the hospital, the director of the device clinic accused me of panicking and assured me I had a Guidant ICD. She was right, but my surgeon, and many of his colleagues, had been persuaded by Medtronic to use its redesigned, thinner lead in me and more than 250,000 Americans—a sales pitch that went on for almost a year after Medtronic and the Food and Drug Administration had received reports of serious malfunctions and several deaths. It took me a long time to get an appointment. It took a full year of transmitting weekly electronic reports from a home monitor to persuade the hospital, the device manufacturer, and my insurance company that the leads were decaying in my chest, evidenced by rapid depletion of the battery.

My only reliable source of information during this year was William H. Maisel, MD, MPH, who was conducting meticulous, original, and humane research on ICDs, the corporations that manufactured them, the federal bureaucrats who approved them, and the people whose hearts were attached to them. His work, much of it published in the *New England Journal of Medicine*, illuminates the medical and ethical complexity of device-based therapies, electronic reporting and remote monitoring, and the overtrafficked, underpoliced intersection of technology and modern medicine.

The fourth and fifth adventures into my vena cava were scheduled for a single day in December 2008. The bad Medtronic lead was to be extracted, along with the Guidant ICD, and a new Medtronic device with the old, reliable Medtronic Quattro lead would be installed—the same model of lead implanted in me in July 2004.

My heart had not changed in size or thickness. I'd experienced no symptoms of HCM. And all of my EKGs were normal—until the day before the scheduled surgery. I was called back after my pre-op tests and told there was something worrisome, something unclear, something very unusual, but after another few minutes, everything was fine.

Everything seemed okay when I awoke from surgery, except for a young doctor at the foot of my bed, riveted by one particular page of my chart. I asked him what was so interesting. He said he was impressed I'd had surgery so soon after having a heart attack. I told him I'd never had a heart attack. He flashed me the page of my chart—the pre-op EKG—which, according to him, proved his point.

Eventually, I tracked down the cardiologist, and the EKG was transmitted to her. She broke out in laughter. This is crazy, she said. It's like somebody mixed up the electrodes for your arms and your legs. It's irrelevant. No doctor could ever mistake this for your EKG.

The last appointment I kept at that hospital was with the surgeon. He was, as always, pleased with his work, pleased with my progress. He was confident the new lead—the old lead, the original lead—wouldn't give me any trouble.

I said, Thank you.

He said, You deserve a break.

I said, That bad lead almost did me in.

He said, It's the first one I regret, the infection. That one could have been prevented.


I was right back where I'd started—almost.

After the first ICD surgery, and with increasing intensity after each surgery, I had noticed that my jugular veins swelled up after intense exercise and hot showers, occasionally accompanied by redness and an unpleasant pressure in my face. My reporting occasioned reassurances that such symptoms were the normal result of raising my blood pressure and body temperature, and I was advised not to let anxiety make me nuts.

In December 2011, I woke up one morning after a few days with swollen jugulars, and I passed out. By then, I had a new cardiologist at a different hospital down the block. He didn't consider my symptoms normal. He asked if Peter could drive me to the Emergency Room, and after many hours there, and many tests, including a CT scan, no one was certain what was wrong with me, but they had ruled out all of the plausible and most alarming possibilities—including blood clots, instability in my heart, pumping problems, and blockage of a major artery or vein.

Over the next few months, I was checked and tested and rechecked and retested. But in relative terms, I was feeling fortunate. My nephew

Nate, who dropped into my life unpredictably for brief, intense periods of contact, was having a terrible time. His heart was failing, and on several occasions, his ICD had fired repeatedly—six or seven times—before he regained consciousness and got to an Emergency Room.

We had dinner during the last week of March 2012. He looked like a fit version of his father. He was heading into exams for his last semester of law school. And he was trying to get his mind around the prospect of a heart transplant, his only real hope for a future. He knew already that I was all for that operation. I was a fan of Nate, and a big fan of life. I'd also known an admirable woman who got a new heart in the 1980s, who lived to become the first heart transplant patient to bear a child. Binny got thirty years out of her second heart. Nate was 27.

Nate told me he didn't to want to die, but lately he was afraid that he would, every day, haunted by his recent near-death episodes. They weren't painful, he said, except for the repeated kicks from his ICD. Instead, he felt exhausted, unable to do what he knew he should do—crawl to a phone. He said the worst of it was how uneventful it was, how insignificant, that his father must have felt what he felt, a deepening loneliness, a sort of resigned, sad sense of knowing, Oh, so this is how it ends.

Before he headed home, he assured me he had an appointment to see his doctors about the next step in the heart transplant process, and a plan to spend the Easter weekend with Mary Ann and her family—two useful prescriptions.

In early April 2012, I was given another echocardiogram, checked out at the device clinic, and I met with the new cardiologist. The swelling of the jugulars was not life-threatening, and it was causing no damage to my heart or the rest of my body. I had developed superior vena cava syndrome, caused by repeated insult to the vein, perhaps involving scar tissue buildup. I'd been rewired too often. Even the best imaging technology is not precise, but it was clear my vein was not as wide as it had been originally—thus the engorgement of the jugular when it was attempting to drain blood during rigorous exercise or very hot showers. In the Emergency Room, I had been warned I might be a candidate for a blood-thinning drug like Coumadin. Instead, my daily baby aspirin, originally prescribed by my sister Mary Ann, was upped to one adult-dose aspirin.

The new cardiologist asked several pointed questions about my genetic diagnosis. I told him I would have the lab send him a copy of the full report. I asked why he was interested.

He said I didn't have progressive heart disease. His reading of the recent echocardiograms had led him to revisit the slight thickening seen in the 2004 echocardiogram. It appeared to him to have been overestimated. You don't have a murmur, he said, a classic HCM symptom, and there is no obstruction at all.

I nodded.

The vena cava symptoms, he said, might be tempered by aspirin (he was right), but the best remedy would be to free up space in that vein.

I nodded.

The battery in my device, he guessed, would have to be replaced in two or three years. He said, How old are you?

Fifty-three, I said.

Next time we see each other, he said, we should talk about removing the ICD.

I said, Taking it out?

He nodded.

I said, No replacement.

He said, Right.

We left it right there.

Ten days later, Nate was attending a black-tie Law Review celebration. His ICD shocked him several times. He never regained consciousness. He was coaxed back toward life on the floor of the ballroom, in an ambulance, and in an emergency room he had visited often. His big, exhausted heart was never again able to sustain a pulse.

In the sad aftermath of Nate's death, in late May, Mary Ann's ICD was emitting beeps at noon—a sign of life in my family. Her battery was wearing out. She checked into a hospital near New Haven, Connecticut, for a routine replacement, a preview of what I might or might not be choosing to do in a few years. The procedure involves opening the pocket, detaching the worn-out device from the leads, attaching the new ICD, and closing up the wound. But Mary Ann's surgeon didn't entirely close the wound, so she left the hospital with a staph infection. Weekly debriding of the infected areas and replacement of several stitches by a plastic surgeon, daily wound cleaning, wearing a sling to prevent any further insult to the surgical site, a six-week antibiotic program, and suspension of normal activities, from swimming to picking up grandchildren, might prevent the staph from infecting the pocket and traveling to her heart.

Help Wanted: What next?

# ELEVEN

# Community and Other Ordinary Miracles

Mara Faulkner

..................................................................................

My dad had an eye disease called retinitis pigmentosa, as do four of my sisters, two nephews, a grandmother and an uncle, numerous cousins, and I. The progress of RP is gradual, unpredictable, and inexorable. There are several kinds, each caused by a different genetic mutation, all leading to the death of the cells that receive light and translate it into colors, shapes, and perspective. Most forms of RP share some physical manifestations: night blindness (as one of my sisters says, at night we're all as blind as a horse's butt) and the gradual narrowing of the visual field until you see a world of dimming colors through two tunnels or soda straws. Though my father was legally blind all my life and almost totally blind when he died, at age 70, blindness was our family secret—the word we didn't say out loud. Until I was 45, I thought I'd lucked out; then an alert doctor saw the telltale pigment encroaching around the edges of my retinas. But not until I began to write my book *Going Blind* did I begin to search for ways to deal with many kinds of blindness, but especially the gradual kind that affects my family—ways that had to be better than our secretiveness and denial.

That's one part of the story. The other part is that I've been a member of a Benedictine monastic community for more than forty years. My Christian faith and the radical choice to live in community have shaped my approach, convincing me that blindness and many other disabilities could be much less disastrous than they often are. You might say I'm looking for miracles.

Retinitis pigmentosa came with my ancestors from Ireland in the steerage compartment of an emigrant ship. My great-grandmother Elizabeth (Bessie) Kelly somehow survived the Great Hunger of the 1840s and, around 1855, came to America with her family, eventually marrying and settling in western Minnesota. She passed RP on to her daughter, Julia Maloney Faulkner, who passed it on to my dad, Dennis Faulkner. In our family, the genes that cause the disease ride on the X chromosome. My father couldn't pass this disease on to sons; but each daughter had a 50-50 chance of having it, and we're all carriers. My dad had one son and six daughters—bad luck for us, the Irish might say. Because RP is progressive but unpredictable, most people who have it live in a twilight zone between sight and blindness, without the simple, devastating clarity of people who are blind from birth or who are blinded suddenly by an accident.

I think I understand what Bessie Kelly felt as she left her green island, with its soft rains and lark songs, to come to the harshness of western Minnesota. I'm on an emigrant ship, too, and I can feel it tossing under me. One part of me is back in the sun-drenched, seeing world. My central vision is still pretty good. I can see leaves just beginning to turn red on our maple tree and the subtle shifts of emotion on my students' faces. I can read normal-size print and was even able to renew my driver's license, though I don't drive at night or in freeway traffic. Most of the time I stand with the see-ers; I still find in myself some attitudes, bred in me by my family and our culture, that make me as blind to blindness as many of the sharp-eyed people around me.

But in the twenty-five years since I discovered I, too, have RP, I've learned something about the fog-shrouded country ahead of me. Like the emigrant Irish, people who've already made the crossing from sight to blindness send messages back to those of us who are just setting out. To find some of those messages, I researched my own family, starting with the letters my dad wrote to my mom during their long courtship. I've had conversations with my family, moving out from this center in ever widening circles to understand both the facts and the feelings that define blindness.

Research gives essential knowledge. Over the years I've gained another kind, based on experience. My visual world is gradually narrowing, as pigment edges in on all sides of my retinas. People and objects to the left, to the right, and underfoot simply disappear; I don't see the people carrying red wine at a crowded party, my students waving futile hands on the far edges of the classroom, or hapless cats, dogs, or toddlers at my feet. I have to be constantly alert to this hard-edged world with its treacherous stairs and booby-trapped floors. I curse architects who thoughtlessly design houses with living rooms two inches lower or higher than adjacent dining rooms, so that blind people will reliably trip. Reading, my beloved obsession, is becoming more and more difficult, not because I can't see the words, but because I have to constantly move my head, tracking each line of print like a little kid with her first reader. Such reading is slow and exhausting. These *cans* and *can'ts* bring a range of feelings that I've read about, seen in my family members, and now often experience myself—dismay at how much of my attention and intelligence have to go into being careful, embarrassment when I'm not careful enough and crash into poles and people, sickening fear, and, often, rage. My grandmother Bessie Kelly couldn't count on the familiar Irish ways to serve her in this new country. Even the ancient craft of planting potatoes was different in the glacial soil and extreme climate of Minnesota. For me, too, the simplest actions have changed and will continue to change. You'd be amazed at how handy peripheral vision is in a world crazy with corners, cupboard doors, and a jostle of people.

I'm learning. But the ways of this new country are strange to me. I don't know yet what it's like to have colors—emerald, burnt orange, lemon—fade to gray; to have beloved faces disappear so that I have to imagine their smiles, their looks of compassion or grief; to learn to use a white cane, carrying it with me in public, knowing how potent and terrifying a symbol it is to the seeing world. In his poem "The Blind Always Come as Such a Surprise," Ted Kooser describes sighted people in elevators and on city streets shocked into silence by the sight of "a great white porcupine of canes." The words of the seeing are "struck down in midflight by the canes of the blind," which tap "across the bright circles of our ambitions."[1] I'll have to learn to go blind well, as my sister Coreen describes her own current efforts. Already I ask myself, Am I too compliant, too needy, too prickly, too secretive? Will I accept well-meaning but misguided help or churlishly turn it down? Will I be patient and humble enough to explain over and over what I can see and what I can't, to ask for the help I need—the friendly arm, the driver, always, the driver? Most terrifying, how will I keep my intellectual, emotional, and creative worlds from shrinking as my visual field narrows? In spite of my best resolve, will I become afraid to trust, to love, to venture out from stifling safety into new places and experiences?

Although I sometimes feel I'm all alone on this emigrant ship, my deepest conviction tells me the opposite is true. That conviction comes from being a member of the Benedictine order. Community life runs counter to modern American individualism, resting instead on the conviction that we humans don't own anything, not even our own lives, and that we are connected to each other by visible and invisible bonds. We carry each other's burdens and are carried aloft by each other's joys.

Recent DNA studies have uncovered our genetic oneness with each other, not just with our fellow humans, but with chimps and yeast and song sparrows. Antonio Gramsci, an Italian cultural theorist, says that to know a person, "It is imperative at the outset to compile . . . an inventory" of the "infinity of traces" our personal and cultural history leaves on the body and psyche of every one of us.[2] Inherited diseases and disabilities, especially those whose lineage we can reliably trace, are among the surest signs that each of us comes into this world marked by an "infinity of traces," binding us in community with many people, living and dead. RP ties me to my Irish past. Bessie Kelly bequeathed it to my family along with her endurance, resilience, and with life itself, a mixed bag, as is all inheritance. The *we* of Benedictine community extends far beyond the boundaries of our monastery, as we turn listening ears to the voices that come to us from others, asking us to pray for a sick child, a job, a drug-addicted son, peace. The long discipline of monastic life has led me to think of blindness in new ways.

I grew up believing in miracles and praying fervently for them—especially the miracle that would heal my dad's blindness and, even more, his anger and sadness. I listened to and believed the gospel stories of Jesus healing cripples, lepers, and blind men,. The only time my family edged close to the forbidden word *blind* was at night, when we all knelt down to say our prayers. We always ended, "Please make Daddy's eyes better." I'm not sure when I stopped praying for miracles like those in the gospels, but I think I know why. I can't ask God to brush aside the pigment on my eyes, or my sisters' or my nephews', because miracles aren't transferable and can't be replicated. A miracle for my dad, me, even my whole family, would leave the hundred thousand people with RP in the United States untouched by healing hands. This says nothing of the two million people worldwide who have RP, and the thirty-nine million who are blind or visually impaired, 90 percent of them in developing countries where even corrective lenses and cataract surgery must seem like miracles.

The miracles I pray, hope, and work for now all need community. The first are the miracles of healing and adaptation that brilliant, empathetic doctors, scientists, and inventors can bring to the community of people with RP and the whole alphabet of other visual impairments. A second clutch of miracles would make life easier for everyone with disabilities, whether inherited or caused by accidents, microbes, or genetic mutations. These are changes in attitudes, laws, and practices as well as structural innovations that would dismantle barriers and confer dignity. Closely related might be the most dazzling miracle of all—communities that gather around any kind of loss or grief, refusing to look away or turn a blind eye to suffering brothers and sisters.

Some might think that these interventions are too ordinary to be considered miraculous. They *are* ordinary, performed by ordinary people like you and me, yet they are still astounding in their generosity, creativity, and practical compassion. Such interventions could (and some already do) alter the stubborn facts of RP and other kinds of blindness. Much has changed since the dark ages of the 1930s, 1940s, and 1950s, when my dad was trying to find his secret, lonely way to a life filled with meaning. Much has changed, but not enough. The miracles I pray for now are extraordinary departures from the attitudes and actions that still deny the gifts and full humanity of blind people—among them, pity, guilt, passive acceptance, lack of imagination, silence, the hastily averted eye.

Recently I attended a conference, "Vision 2012," sponsored by the Foundation Fighting Blindness, whose goal is to cure or at least slow down the progression of a range of retinal diseases, including RP. For the first time since I heard the name "retinitis pigmentosa" more than fifty years ago, a cure, or at least useful medical intervention, is within reach, if not for me and my sisters, all now in our sixties and seventies, then surely for the next generation. With the help of generous donors and grants, this foundation and many others are funding research and clinical trials around the world to test the efficacy and safety of several different therapies: gene replacement, stem cell transplants, artificial vision created by several kinds of bionic eyes, drugs, and nutritional supplements. A worldwide community of imaginative, pragmatic scientists and doctors is at work. As they are quick to point out, each of these therapies holds its own promise and peril; most are still years away from government approval and widespread, affordable use, even in developed countries. Researchers described the difficulties each therapy presents (including being hugely expensive to develop and test), but none of them are ready to give up, and they urged us to have the same patience and hope that guides their work. On the last day of the conference, someone asked how soon gene therapy for X-linked RP, my family's kind, will go into clinical trials. When a lead doctor answered that it won't be tomorrow, but it won't be five years either, I could feel hearts lift with an eager hope none of us had felt before. Scientists would not use the word *miracle* to describe something as material, as laborious and time-consuming, as genetic research; but that room was filled with people who've heard the diagnosis for themselves or their children: "It's retinitis pigmentosa. There's nothing we can do." For us, hope for gene therapy is all the miracle we need.

Until that day arrives, and for many blind people it never will, the Foundation Fighting Blindness, among other organizations, is working at innovations that will make life as full and free as possible for people who are blind or going blind. Some are still in the planning and dreaming stages. One is a high-tech cane developed by Dr. Amir Amedi of the Hadassah Medical Center in Jerusalem. This ingenious cane, "by processing tones and vibrations, lets the user know what is ahead and to the sides. It also indicates changes in depth—for example, when a person reaches a curb and needs to step down." Since blindness doesn't automatically confer supersensitive hearing and touch, a cane that signals an approaching flight of stairs or a car turning left into your path would come in handy. Even more astounding are Dr. Amedi's efforts to retrain the visual cortex so that blind people will be able to interpret the sounds they hear, with their varying tones, pitches, and lengths, as letters, words, and faces.[3] When I think of my sisters, who now see the faces of their grandsons as milky blurs, I know this reprogramming would seem like—would be—a miracle.

It was miraculous when developers asked blind people what they need and want, and in response developed such devices as smartphone apps and GPSs designed especially for them, so that people like my sister, who is both blind and direction blind, can find her way there and back. While some programs and workshops for the blind are mired in ancient stereotypes that relegate participants to making kindergarten clowns out of cotton balls and popsicle sticks, others are both responsive and challenging, recognizing that blind people are people first of all, with diverse talents, interests, needs, and aspirations. These programs offer training in the visual arts, fashion design, sports, and many levels of self-defense, from down-and-dirty street fighting to tae kwon do.

Other helpful accommodations involve city planning and reliable, timely, affordable public transportation. It is possible, though still rare, to design cities and transportation systems to benefit not only the blind but anyone who can't use a car to get to where the jobs are, often in far-flung suburbs. This simple change might not seem like a miracle in a city with public transportation, like New York or Boston or San Francisco, but it would in Minneapolis, where one of my nephews lives. He has degrees in architecture and city planning and development. In his most recent job search he has run up against two related barriers: jobs for which he is qualified are located in those distant suburbs where city transportation doesn't go, and job applications routinely ask if he has a valid Minnesota driver's license. Chad has never seen well enough to drive, and that question, which he has to answer on electronic applications, filters him out (along with undocumented immigrants and felons), in spite of his qualifications and experience and the fact that many of these jobs don't involve driving.

Much has changed in the attitude of the blind people striding boldly into the world, in their families, and in the world that greets them. Some believe the stigma long attached to blindness is gone, at least in the United States. I see heartening signs that this is true—and other signs telling me we still have far to go.

First the hopeful signs. At conferences, at workshops, and online, blind people urge each other not to let anything stand in the way of their aspirations, especially not the low expectations and rejections they will inevitably meet. At the Vision 2012 conference, Eric Weihenmayer, who has climbed every tall peak in the world, urged listeners to "feed on frustrations and setbacks and emerge on the other side, not just damaged as little as possible, but better, stronger," more able to serve others.[4] He didn't expect the five hundred people in that room, sighted or blind, to climb Mount Everest, but rather to envision the life they want, and go after it. This is the same message Tim Cordes, one of only a few blind physicians in the United States, delivered to the National Federation of the Blind in 2010: "I look forward to a world where people with disabilities do what they can and what they want, and it's not exciting or different." That will be the day when a top student like Cordes doesn't get turned down by all but one of the medical schools he applies to. He got into the University of Wisconsin–Madison thanks to the backing of one supporter. School administrators feared that accommodations for Cordes would be too expensive and, incredibly, that "the Association of American Medical Colleges might frown on a school admitting a student who couldn't see."[5]

Blind people don't just talk about living boldly. They do it every day, participating in all kinds of sports at every age; taking on challenges like hiking all two thousand rugged miles of the Appalachian Trail; entering almost every profession—from teaching to astrophysics and high finance. Most important, they live a normal family life. Ryan Knighton recently published a funny, irreverent book called *C'Mon Papa: Dispatches from a Dad in the Dark.*[6] Through books like this one, through blogs and websites, blind people and their families offer each other encouragement, helpful hints, and, in Knighton's case, some how-not-to's, such as how not to lose your little daughter in the snow. Rather than overprotecting their children, many parents now follow the advice Tim Cordes's mother gave herself when the doctors listed all the things her son would never be able to do: "[I] decided to forget everything they told me."[7] You'll find parents cheering for their blind children at wrestling meets, cross-country races, and black belt competitions. Many people are openly, unapologetically blind, seeing no reason for the denial and secrecy my father used as a shield for his pride and his wounded spirit. Even as they claim their blindness, they refuse to be blotted out by it or relegated to the sidelines of life, which for most blind people is their greatest fear. The provisions and legal muscle of the Americans with Disabilities Act, passed in 1990 and amended in 2008, have helped to make some of these advances possible. Without the ADA, I'm pretty sure that the University of Wisconsin would not have provided Tim Cordes with a machine that creates raised drawings of images, which he reads like a Braille text, and human "visual describers" to read texts that reading machines can't decipher.

Still, other blind people and I continue to come up against barriers in the form of spoken and unspoken assumptions. Recently a friend who knows I'm going blind said, "If I ever lose my eyesight, they might just as well shoot me." There it is—the old, destructive conviction that blindness is worse than death. By implication, the blind person (and maybe the world) would be better off if he or she (or I) had never been born. If a blind person hears and believes that death sentence, blindness can spread like a terrible stain from eyes to mind to body to heart, until it becomes the self. I have been asked if I have conflicted feelings about having inherited retinitis pigmentosa from my father. That question had never occurred to me. I'm not sure what feelings the questioners had in mind—regret? gratitude? resentment? Though my feelings are mixed, resentment isn't one of them. I wish my beloved sisters and nephews and I didn't have RP. I hope and pray for a treatment or cure, and when they find one, I'll be in line. I regret my dad's unhappiness and his inability to teach us, his children, to "be blind well." But I don't for a second regret my life, the sheer, astonishing fact of existence.

Related to the notion that blindness is worse than death is another piece of misinformation: without the 70 percent of information sighted people take in through their eyes, the sensual world is closed like a stone. When I told another friend about the man who hiked the Appalachian Trail, she asked, "Why would he want to do that? He can't see anything." True enough, but there's the smell of pine sap rising, the bracing ozone smell of coming rain, the sounds of the rapids and the birds, the wind, the cold, the good pain of a body stretched to its limit.

These are anecdotes, you might say, remnants from a benighted past. Maybe they are. What is not an anecdote, though I wish it were, is the unemployment rate for blind people in the United States. Although it's hard to find exact figures, most organizations of the blind report that almost two-thirds of blind people are unemployed, a scandal and a tragedy for each person and for the community that would benefit from their contributions. This stubborn figure has been stuck near there for several decades, in spite of the ADA, hundreds of technological miracles, and the grit and resourcefulness of many blind people and their good allies.

I am a few months away from the end of a long, exhilarating teaching career. My work has been a gift, a calling, a contribution to several generations of students; what Stephen Kuusisto, author of *Planet of the Blind*, calls "the joyous striving that necessarily defines a strong and good life."[8] I want that gift for every person, including every blind person. I wonder what I would do if I were 22 years old again and my RP had progressed enough to make me legally blind. Would I preserve the illusion of sight as long as possible, to get the job and keep it, earning tenure and the security it confers? Or would I walk confidently into the interview with my white cane or guide dog, trusting that the school would judge me on my merits and, if I were the best candidate, hire me and make the "reasonable adaptations" mandated by the ADA? Honestly, I don't know what I'd do, or what counsel I should give my nephews, who still have some usable vision. Most blind and visually impaired people aren't Tim Cordes. We're mostly ordinary people with ordinary intelligence and gifts, wanting to earn a living for ourselves and our families or, in my case, my Benedictine community. We shouldn't have to be brilliant. Nor should we have to be belligerent, cringing, or wildly courageous to get necessary accommodations such as a classroom aide, a grader for math papers, or a proctor for tests. So, I'm looking for another miracle—a shift in attitudes and actions that require more of the sighted community than admiration and praise for "extraordinary" blind people. It isn't the job of disabled people to inspire the rest of us slackers! I want the miracle of an unemployment rate for the blind that matches that of the rest of the population—as of mid-2012, 7.8 percent.

Humanizing community support doesn't always materialize, that is, become real, fleshy. Almost every account of blindness I've read, even those written by people who have successful and meaningful lives, tell of times of deepest despair, fear, and anger. My sisters and nephews have told me about the inner darkness that sometimes overwhelms them. The despair comes partly from blindness itself, especially progressive blindness that brings with it unpredictable waves of loss deep enough to swamp the stoutest heart. But it also comes from the recoil of well-meaning people. It's easy to stare at blind people and even easier to turn a blind eye to them because they can't see you doing it. But my sisters and nephews can feel that blind eye sliding away when they find themselves sitting alone with their canes at a crowded party or family gathering, "in the shadows," my sister says, as the joyful chaos of life passes her by.

We in the developed West are inordinately afraid to look at suffering, disease, and imperfection. In fact, we have a phrase to describe our reluctance and social psychologists to measure how quickly it sets in. *Compassion fatigue*, we call it, and, according to *New York Times* journalist Nicholas Kristof, who has laid before American eyes some of the greatest tragedies of our times, this fatigue overwhelms his readers very quickly. He says that "the point at which we begin to show fatigue is when the number of victims reaches two."[9] Compassion fatigue is real; I've felt it. But this comforting phrase gives us an easy out, so that we don't have to look with empathy and respect at the third starving child, the third blind person tapping her cane across our unbroken lives.

When support materializes, it's a miracle as stunning as the blind man touched by Jesus, dancing down the streets of Nazareth shouting, "I was blind, and now I see." As stunning as a blind woman, my sister, running pell-mell across the playground, hand in hand with her grandsons, drunk on spring and laughter and freedom. To describe the miracle of support, I have to step aside from inherited blindness and look at mortality itself, that trait passed reliably from every parent to every child. My family hasn't faced more tragedies than the next, but we've had our share, including the deaths of babies, not from genetically transmitted diseases, but from accidents of birth and icy roads.

A few years ago, we were all happily anticipating the birth of my niece's twin daughters. A few days before they were due to be born and only a week after the doctor heard two hearts beating lustily, one baby died. Mariah was a full-term, healthy baby with no discernible cause of death. A community linked by compassion that transcends genetic pathways gathered around my niece and her family. They knew they couldn't change what had happened, they knew no words or actions could take away the pain, but they didn't look away. Instead, they formed a protective circle, all of them willing to carry part of the burden. A group of volunteer photographers called Now I Lay Me Down came to the hospital to take pictures of Mariah and her family; they made a video to help us remember this small breeze that blew through our lives. At the memorial service for Mariah, my niece's violin students, mostly little kids, played "Sleep my child and peace attend thee, / All through the night."

Just a year earlier, another of my nieces lost her father (my brother) and her baby son in a car accident. After the accident, Jessie kept a blog, The Encouragement of Light, to stay in touch with faraway family members and friends, but also to reach out to others who have lost children, putting the inexpressible into eloquent words. As she and her husband traveled deep into "the strange country of grief," her blog gathered readers and respondents. Later, pregnant again, she wrote, "It is different now. So many of the people I have contact with have lost a child, and their stories are part of me now. The fragility of life is always in my awareness. But some days, I really wish for the innocence of not knowing grief."

So do we all. But as Jessie and Michael came to understand, the cost of such unknowing is the loneliness and despair of our brothers and sisters. Stephen Kuusisto writes about the opposite of isolation and despair in his essay "The Beauty Myth." He took a group of students to London to see some of the artifacts they'd read about in his Disabilities Studies course at Ohio State University. They'd studied the development in Victorian England of institutions that shut disabled people away from the gaze of fashionable café crowds and tea drinkers. Kuusisto has been nearly blind all his life, uses a cane, and usually has his guide dog with him. On this trip, he asked his students to be "sighted guides":

> I taught them to guide me with their outstretched elbows, how to
> help me locate the steps of the Underground, how to find curbs. We
> traveled as a group and worked our way through the rush hour
> throngs of Trafalgar Square. Each student had a chance to guide her
> blind professor in an unfamiliar city. Each also had the opportunity
> to describe what she was seeing, whether we were on the street or
> deep inside the Victoria and Albert Museum.[10]

That magical high-tech cane might have guided Kuusisto as surely as his students. He might have gained independence, but he and his students might have lost the warmth of human touch and community.

Recently, my almost blind sister got her first guide dog from Leader Dogs for the Blind, in Rochester Hills, Michigan. Kaiah is a whip-smart German shepherd with warm eyes and fawn ears. Every day they emerge from the safety of their house to take on the airy, dangerous world. They walk to spots where the city streets end and the mountain trail, soft with pine needles and rich with the scents of animals and wind, snakes up into the Montana foothills. With utmost kindness and patience, Kaiah follows each command—"Kaiah, find the door, the curb, the gate"—never mocking her friend with all she can't see, never embarrassed to be seen in the rigid yellow harness emblazoned with the words "Do not pet me, I'm working." She's not self-important; she's just doing her job with her senses and intelligence alert. Sometimes they're both afraid—of escalators, airplanes, crowds of people. But they don't retreat from the world, and two fears enable one brave act after another. You might think this is a solitary miracle—one woman, one dog, taking on the world. But a large, mostly invisible community made this lovely partnership possible. Imaginative people established the first guide dog center in the United States in the 1920s. The financial support of various groups and individuals makes dogs potentially available to all blind people, regardless of means. Puppy trainers, dog trainers, and guides in several centers around the country match up dogs and people and train them to work as a team. This partnership asks as much of the human as it does of the dog! The trainers work them hard for several weeks: up at 6:30 to "park" their dogs, and then out into city traffic. The dogs are part of the community and make it grow. People who would have passed my sister by, with a shifty, side-long look at her cane, now stop to talk, drawn to this beautiful dog and her human friend. Kaiah has led Coreen out of the shadows, something her cane could not do.

I've always been afraid of the dark. When I was a kid, I hated the unlighted front yard of our house, the hiding place under the porch, the darkness outside my bedroom window, where I was sure black panthers lurked—on the outskirts of Mandan, North Dakota. And yet, my dawning understanding of blindness has led me farther and farther away from Western culture's obsession with light as the only good and safe place to be, an obsession enshrined in Judeo-Christian imagery where God dwells "in unapproachable light." I find myself turning sometimes from that oppressive light to darkness. In his poem "Night," the seventeenth-century Welsh-English mystic Henry Vaughan writes:

There is in God, some say,
A deep but dazzling darkness, as men here

Say it is late and dusky, because they
See not all clear.[11]

In that deep darkness is all I don't know and have to learn about being blind in the years ahead. Because I'm still on that rocking boat somewhere between the world I know and the new world, which is both familiar and utterly strange, I want to learn to be a better sighted guide, as alert and kind as Kaiah, happy to be walking with my sisters, my nephews, with all the blind. Kaiah knows what I still have to learn—that "pity is 100 percent curable." This motto of the Gillette Children's Hospital in Minneapolis suggests that pity is a disease or a disability that disables both the pitier and the pitied. It's contagious and transmittable from generation to generation in the societal DNA. It is the opposite of communal bonds, which rest on mutual respect and the equal dignity of each member.

Whether we can see or not, if we have the courage to leave this light-drenched land for the dazzling darkness, we'll find loss and grief, despair, ignorance, and even cruelty. But we'll also find the unquenchable hope I've been describing that only community makes possible. Like Stephen Kuusisto, we'll find that "roses grow on the sheer banks of the sea cliff."[12]

## PART THREE

# Passing Down

*Genetics and Family*

# TWELVE

# String Theory, or How One Family Listens through Deafness

Jennifer Rosner

························································································

After our daughter, Sophia, was diagnosed deaf, my father told me the story of my deaf great-great-aunt, Bayla, and the string she used to "hear" in the night.[1] My father heard the story from his mother, who must have heard it from hers. In the 1890s, in an Austrian village, a small community lived together. In this shtetl of Jews (or so the story goes), there was a shoemaker, a water carrier, a jeweler, a milkman. There was a rabbi. There were students and scholars, husbands and wives. Among them, there was also a deaf girl—my great-great-aunt, Bayla.

When Bayla grew into a young woman, she married and had a baby of her own. A little girl. In the daylight hours, Bayla watched her daughter carefully, scouring her expressions to assess her needs. But daylight was followed by night's darkness. How would Bayla know if the baby needed her when she was sleeping? Bayla came up with a plan: before she went to sleep each night, she tied a string from her own wrist to her baby's. Looped underneath Bayla's dressing gown sleeve, the string trailed from the bed, across the floor, through the slats of the crib, to end in a floppy bow around her baby's wrist. Whenever the baby fussed or cried, her movement pulled the string taut. Bayla felt the tug and woke to care for her. That string was Bayla's way of hearing her baby in the dark.

Listening to Bayla's story, I remembered how much I'd longed to forge a strong bond with my children, even before they were born. My connections with my own mother had been intermittent and shaky. Growing up,

I felt my mother tuned me out. She had serious hearing problems of her own, apparently caused by mastoid infections and botched surgeries in childhood. But in my mother's case, the communication issues went deeper. Her father had run off when she was 10 years old, and her mother had a limited capacity to attend to her. My mother knew only broken chains—she had no string to tie from her wrist to mine. I was determined to reverse that course with my own children.

When Sophia was born, my husband, Bill, and I snuggled together with her in the hospital room, and I felt *sure*. Sure that we were going to be connected, healthy, and well. Within hours, that certainty was shaken. A technician wheeled in a computer cart to run the universal newborn hearing test on Sophia. Stepping gingerly around the computer wires, she peered at the monitor, punching on the keyboard as she scrutinized the data. I couldn't take my eyes off Sophia. She was all olive, rose, and brilliant slate gray, her features still mushy like a newborn pup's. But when I glanced over at Bill, I was surprised to find him ashen-faced. He was asking the technician about the numbers that appeared on the screen, his gaze darting back and forth between Sophia and the monitor.

"What are the typical scores for this test?" Bill asked.

"Usually in the hundreds," the technician admitted.

I sat up and gaped at the tiny white numbers on the screen: 4 . . . 7 . . . 3.

The technician shifted her weight. It was clear she had no idea what to say. She stuffed the tangle of cords on a low tray, switched off the computer screen, and guided the computer cart, backward, out of the room.

Something was wrong with Sophia's hearing. Though we were discharged from the hospital the next day, we were scheduled to bring Sophia back for further audiological testing.

I immediately dialed my parents, and my father picked up the call. Did he think Mom's hearing loss was genetic? Might there be any reason to think deafness ran in our family lines? Denial is a powerful force; my father didn't say a word about the deaf relatives in our family. He suggested, instead, that Sophia's hearing loss was caused by excess fluid in her ears and that her hearing loss would clear up on its own.

Only later would I learn of the deaf ancestors peppering my family tree, each one's name duly marked with an asterisk on a sprawling handwritten chart bordered by a disheartening key: * = deaf and dumb. Only later would I discover that my father brought this very chart to a geneticist back before he and my mother started planning a family and was told that, due to the patterning on the chart, genetic deafness was unlikely to recur on our particular branch. (This same response was actually echoed by *our* geneticist, forty years later.)

Only later would I hear the story of the string, wrist to wrist. My ancestor's keen determination to hear her child.

Follow-up brainstem response tests showed Sophia to have severe sensorineural hearing loss. The audiologist held up a piece of graph paper spattered with drawings of sounds: a barking dog, a ringing telephone, an airplane, a piano. A jumble of letters—representing the sounds of speech—clustered inside a banana-shaped outline. A sloping line, like a mountain sketch, overlaid the drawings. The audiologist explained that the sloping line showed the range of frequencies and decibels Sophia could hear. There was almost nothing in the "speech banana."

I had spent the last two weeks running homespun hearing tests, and while Sophia had, at times, seemed to respond to the clap of our hands or the bark of our dog, she'd never so much as blinked in response to the sounds of our voices. If she couldn't hear spoken language, she herself might never speak. As I sat there in the audiologist's office, my dreams of a child's chatter and song began to wither inside.

"Does this mean she can't hear us talk?" I asked the audiologist hoarsely.

"She may hear some speech sounds—the wide open vowel sounds— but not much else." She explained that the typical inner ear has more than fifteen thousand tiny hair cells that convert sound waves into neural signals. In Sophia's case, many of these hair cells were probably broken, bent, or missing. As the audiologist spoke, she passed me a binder thick with information about deafness.

I was shaking before I felt it. I rocked Sophia, rocked her back and forth in my arms. She was just 2 weeks old, her tiny hands still transparent.

*What would life be like for her?*

*How were we going to communicate?*

The audiologist continued to fill us in, telling us about hearing aids, about Sign Language, about the cultural divide between Deaf and Oral schools. She told us that deafness was sometimes syndromic and that

Sophia would need further tests to rule out eye, kidney, and heart disorders. We should schedule a genetics consultation, too, as congenital deafness could be hereditary. It was hard to take it all in.

Bill and I were catapulted into parenthood like all new parents, but we had the added work of researching sensorineural hearing loss. I had a PhD in philosophy; Bill was a child advocacy lawyer. We both knew how to dig for information, and we got busy. In those early days of research, we learned that 3 in 1,000 children in the United States are born with some degree of hearing loss. Ninety-five percent of these children are born to hearing parents. We also learned that genetic deafness has a higher occurrence among Ashkenazi Jews (our heritage). As regards disease frequency, nonsyndromic congenital deafness (DFNB1) occurs in 1 in 1,700 Ashkenazi Jews, compared with 1 in 7,000 in the general population. As regards carrier frequency, the rate is 1 in 20 to 25 among Ashkenazi Jews, compared with 1 in 35 among the larger U.S. Caucasian population.[2]

In researching the communication options, we discovered a deep political divide between those who believe that, if possible, deaf people should make use of hearing aids or cochlear implants to help them hear and speak, and those who believe that deaf people should embrace Deaf culture and communicate primarily by using Sign Language. Some in the Oral camp consider it a moral obligation to give babies (when practicable) access to the sounds of speech and the larger, hearing world. Others, on the side of Deaf culture, consider such assimilation a travesty, arguing that deaf babies are "altered" by technology, "mutilated" by surgeons, and stripped of their rightful membership in the Deaf community.

We felt like we'd stumbled into a minefield: any answer carried a prejudicial judgment. Was deafness a disability? Did it constitute an essential identity? We wanted what was best for our baby, for our family. It was infuriating to feel like our decisions for Sophia might be judged by others, and in such polarized terms.

Yet, we had to make decisions about how to communicate with Sophia, and fast. Sophia, like all babies, needed exposure to language. The audiologist presented us with the options: we could take a spoken language approach, we could immerse ourselves in American Sign Language, or we could try *both*: we could embark on a "total communication" approach by aiding Sophia with hearing aids and speaking, while signing with her at the same time. As we grappled with the intricacies of each option, we had the added burden of advice from dozens of unsought "consultants." Everyone around us had an opinion—whether we'd asked for one or not. These opinions were often given without considering the unique features of our family, the particularities of Sophia's hearing loss, or our communication style and desires. Not to mention our very personal feelings about life's opportunities and familial intimacy.

At a work-related party, Bill explained to some co-workers that we were considering high-powered hearing aids for Sophia.

"Why don't you just let her be who she is?" The man standing to my left was admiring Sophia, but he was speaking to me.

"What?" I asked, taken aback.

"Why don't you let Sophia be who she is?" he repeated.

"Who *is* she?" I looked at Sophia. She was 2 months old. Did she already have a fixed identity?

"She's deaf," he answered. "She was born without access to sound. Why not let her live that way?"

DEAF. That couldn't be *who Sophia is*, could it? My philosopher's hackles were up. Just as I started to object, he excused himself to chase a tray of stuffed mushrooms.

In our hearts, we wanted Sophia to hear, and we wanted to communicate with her in our native (spoken) language if at all possible. While we intended to learn ASL so that Sophia could communicate with Deaf people she met and with whom she shared many experiences, I hoped that this unfamiliar second language would not become our only means of communication—certainly not while I was in the early stages of learning it. I felt this way both for Sophia's sake and for mine. If we relied solely on ASL, we would need an interpreter, fluent in the language, for the critical early months of Sophia's life—someone living with us, involved with every word that passed from one of us to Sophia. I did not want an ASL interpreter or any other person mediating my relationship with my baby. I wanted to connect with her more intimately, in my native language. As the mother of a new baby—as a new mother who had grown up with frayed connections—my primary concern was for closeness and intimacy.

A further issue had to do with opportunity. We feared that Sophia's opportunities would be narrowed if her communication was limited to Sign. Only 0.02 percent of people in the United States are fluent in American Sign Language. If Sophia could gain access to spoken language, she could communicate with the much larger hearing world.

It is extremely hard to make a decision for a prelingual child. Sophia couldn't tell us what she wanted, and we didn't have time to wait. Delays in stimulating neural pathways from her auditory nerve to her brain would reduce Sophia's chance of ever gaining auditory access. If we were going to give the spoken language approach a try, the time was now. We decided: we would start with hearing technology and auditory training and revise this decision later, as needed.

Sophia was fitted with digital hearing aids when she was 3 months old. The aids were huge and flopped off her tiny ears. But almost at once, she began to pay attention to our voices and to vocalize.

At 11 months, she uttered her first word.

"Up!" she asserted, and I whirled her *up*!

I called everyone we knew. Sophia's first word was *up*. She was an optimist!

Sophia's spoken language capacities exploded, and we began to feel confident about the choice we'd made. With the support of an auditory-oral teacher of the deaf, Sophia learned to attend to sound with great perception: before long, she was hearing and speaking. As a family of three, we were closely bonded.

We still didn't know the cause of Sophia's deafness. According to the geneticist we consulted, my family history did not clearly indicate genetic deafness. We could test Sophia's blood—our insurance would not cover a test of my blood, or Bill's—but the test required a large draw. Sophia was very low weight, and we were worried about her growth. We declined the test for Sophia. We felt we were handling her deafness well, and if our next child was also deaf, we'd know how to proceed.

Months later, when I became pregnant, my doctor offered to test my amniotic fluid for the genetic mutations that cause deafness. This information, it was suggested, would help us decide whether or not to continue with the pregnancy. We declined. We wouldn't act on the information either way.

Still, I worried. In the hospital, I made the nurses promise that they wouldn't perform any tests after the baby was born unless Bill or I were there. Especially not the newborn hearing test.

Juliet was born in September, just after Sophia's third birthday. I was exhausted after labor, and slept deeply. I woke to see a nurse standing over the hospital bassinette. Juliet was swaddled and sleeping, electrodes stuck to her head and blue gooey stuff oozing into her hairline. I hastily attempted a sit-up, a crampy pain clenching my womb. Before the nurse could explain, I reminded her what we'd said about tests. *Not without our permission. Not without one of us there. And awake.*

The nurse gave me a measured look and said, "I'm just repeating the hearing test to be sure of the result."

"*Repeating?*"

"Yes. Your husband was with me earlier when I performed it the first time. You were sleeping."

The white drawstring on her hospital pants had the blue goo on it, too.

Flushing hot, I propped a pillow behind me and leaned back against its cool side. Bill had just been here, but didn't tell me. "She failed it?"

"Yes. But I thought I'd run it once more to be sure."

The crampy pain was back and I shifted positions. "Is she failing it again now?"

"Yes."

I had told myself I'd know what to do if Juliet was born deaf, but it turns out, I didn't. One day, I laid Juliet down on a satin-edged baby blanket in the center of the living room floor. The smell of zinc oxide from her diaper cream mingled with our wool rug, rough beneath my knees. I tried to meet Juliet's gaze, but her eyes didn't register mine. She looked in my direction, but past me, through me, toward the sunlit window. It struck me that she hadn't met my gaze once yet. I leaned over her, putting my face right over hers, but she still didn't look at me. She began to arch her back.

Juliet's spine formed a perfect bridge, her weight balanced on the soft spot of her head. I rearranged her, bending her knees up toward her belly to force her into a concave posture, but as soon as I took my hands away, she reassumed the arched position. I backed away—was she arching to get distance from me?—but she stayed in her backbend, the light from the window bouncing off her upside-down chin. She couldn't hear me. And she wasn't looking at me, either. Whatever competence I felt as a mother the second time around—I was skilled now at newborn feeding, diapering, bathing, and swaddling—was undermined by Juliet's inexplicable arching, her distant gaze, and the unknown degree of her hearing loss.

In the days before we met with the audiologist who would test Juliet, I tried, at home, to gauge whether Juliet could hear. One morning, I draped her over my shoulder and sang to her from a mixed-up, past repertoire of show tunes. I landed on Freddy's song from *My Fair Lady*—I knew all the words. I could picture the freckled, redheaded boy who sang the song in the production I was in as a teenager, back when I dreamed of becoming a singer. "I have often walked down this street before;/But the pavement always stayed beneath my feet before./All at once am I/Several stories high,/Knowing I'm on the street where you live." The boy belted it out full-volume and now, so did I. I didn't stop for Bill's footsteps down the stairs, and I continued singing as he looked on.

"I don't think she can hear anything," Bill said when I was finished. The brainstem test that would tell the exact degree of Juliet's hearing loss was still a week away, but Bill felt certain Juliet was completely deaf.

He was right. According to the audiologist, Juliet couldn't hear anything at all. Not a running lawn mower, not a revving jet engine. Hearing aids, no matter how powerful, wouldn't enable her to hear spoken language. The reason Juliet had been arching her back, the audiologist surmised, was to try to look behind herself. Juliet was "profoundly" deaf.

I had half-expected deafness again, but I had *not* expected complete deafness—deafness that even hearing aids couldn't touch. We couldn't be sure that a spoken language approach was even an *option* for Juliet. What had worked with Sophia might not be a choice for Juliet.

In fact, we were told that if Juliet were to hear at all, she would need cochlear implants.

First, however, Juliet would need an MRI to determine whether her cochlea was structurally intact for a cochlear implant. That test could be performed when she was 6 months old. If she was a candidate, she could undergo cochlear implant surgery at age 1. Then, with extensive training, she could learn to hear (albeit electronically) and later, to speak. We embarked on Sign Language immediately with Juliet and held out hope that, in time, we could add the capacity for hearing and spoken language.

Back once again in the Hearing-Deaf divide, we found that many of the considerations we'd faced in Sophia's case were the same in Juliet's: only 0.02 percent of people in the United States are fluent in ASL, compared with the 99.98 percent of people who use spoken language. If we could raise Juliet in the hearing world, we would be giving her the chance to communicate with the larger, hearing population. And not only with the larger population, but with *Sophia*. We wanted Sophia and Juliet to develop strong bonds growing up. Sophia was acquiring some Sign, but it was already clear that she was learning—and flourishing—as a hearing, speaking child. Likewise, Bill and I each wanted our own relationship, our own intimacy, with Juliet, and we felt that we could better achieve it by communicating in our native tongue.

The idea that deaf children "belong in" and flourish better in the Deaf community makes a great deal of sense to me if the child's extended family is Deaf and if their native language is Sign. It makes less sense if a child's parents and family are hearing and speaking and, consequently, outside the Deaf community. Weighing this decision for Juliet, I understood that in the past, deaf children often found solace in a separate, Deaf community. But I felt it was imperative for *our* family to find a way to share in the same community, one way or another.

After a variety of tests, we received a call from an ear surgeon at Children's Hospital. Juliet's MRI showed her cochlea to be structurally intact; this meant that implant surgery was an option. Electrodes could be strung along the curvature of Juliet's inner ear to do the work of her broken, bent, or missing cilia. With the implants, she could "hear" through an electronic process; she could have access to language and sound.

The surgeon also had results from Juliet's blood work.

"Juliet's hearing loss is genetic," she said. "She has two deletions at the Connexin 26 gene site—35 del G and 167 del T."

I scribbled this down.

"Recessive gene mutations, one each from your husband and you. In all likelihood, Sophia has these too."

That afternoon, I went to my computer to look up Connexin 26. Online, I found an article about the genetics of deafness. Ordinarily, there are pairs at the gene site. Genetic mutations can cause deletions, with the result that there are singletons where these pairs should be. The Connexin 26 mutation is autosomal recessive—with two carrier parents, there is a 25 percent chance for any given child to inherit it.

I gathered up the pages of my family chart, scattered across my desk. I stared at all the names, running this way and that, scribbled crosswise, scampering up and down branches of the family tree. Singletons, in desperate search of the missing halves that might make them whole. Make them hear. Make them heard.

When I told my parents about the recessive genes, my mother's response was to express relief that Bill's family shared half the responsibility for the girls' deafness. My brother and his girlfriend immediately scheduled themselves for genetic testing, with the implication that they would sooner refrain from having children than risk the chance of having children who couldn't hear. I knew, then, that the deafness figuring in my larger family far exceeded the literal damages suffered by my girls' organs of hearing, and *that* deafness was far more disheartening.


At age 1, Juliet received her first cochlear implant. After several months of auditory training to help her make meaning out of the sounds she was newly hearing, she began to vocalize, to turn her head to sound, and finally, to speak. We were thrilled!

The day Juliet uttered her first clear word—*milk*—we drank and added milk to everything! At the market, I bought vanilla milk, coconut milk, maple milk, even strawberry milk—just to prompt that precious word from her lips again and again.

Soon Juliet was imitating everything she heard, chattering up a storm.

But there were times that Juliet did not want her "sound" on. I'd place the implant processor on her (there is an external part, which sends auditory information to the internal, implanted part). She'd glower at me and bat it off with her hand. I let her putter around the house, then. Soundlessly.

Without the processor on, Juliet played in silence. She built a wooden tower, then knocked it down—*without* the kaboom. She typed on the keyboard of her play computer—*without* the click click click. I took her out to the swing set and let her swing back and forth *without* the creak creak of the metal chains. What was it like without sound?

Juliet was happy enough to let us put on her processor each morning, but if a restaurant was too loud or she was tired, *yank*—off it would come—and I'd rush to catch her thousand-dollar "ear" before she plunked it on a table smeared with ketchup or blue cheese dressing. Just as quickly, I raced to settle Juliet into a comfortable position, because with the blessing of quiet and the downward flicker of her eyelids, she'd be asleep in seconds.

Juliet ripped her processor off at the sight of the blender or the vacuum cleaner. She put it back on when she spied a favorite book or video. On and off went her sound—off for the bath, on for the play date, off to skitter through the sprinkler, back on for singing with cupcakes. Why should she endure babies crying at the pediatrician's office if she didn't have to? Why not miss the jackhammer blast when the car was stuck in traffic? Juliet sensed much of what was happening through vibrations, anyway. Whether her sound was on or off, she knew when someone walked into the room, and she could read a face like a book.

Juliet's favorite face was Sophia's, and their play didn't need hearing. Sophia would put herself nose to nose with Juliet, locking eyes to secure her attention. Then Sophia led Juliet in play: they toted their baby dolls from room to room, they gave each other foot baths with bowls of warm water, they emptied every drawer in the house for inspection. They put on music, whether or not they could hear it—sometimes they *both* had their sound off—then donned butterfly wings to dance to its vibrations. So what if Bill and I might ourselves go deaf with the volume turned way up? They *felt* the music. They were having fun together, and they understood each other perfectly.

But the girls needed their technology *on* to hear language and to continue developing speech. After a break, we'd put their sound back on.

Sophia and Juliet are part of am emerging hybrid: the hearing deaf. When they wear their sound technology, they can hear and speak, and they have developed incredible language skills and beautiful speech. When they take their technology off, they are deaf (though they continue to understand the world through the categories of spoken language). With the advances in technology and newborn hearing testing, this may be one future of deafness. I often wonder how the fault lines of the Hearing-Deaf divide will shift in response to it.

Even with the differences in the degree of their deafness and the technology each uses—my girls share more between them than either Bill or I are able to share with them. I am happy that they have each other. As a family, we are more connected now than I ever dreamed, and Sophia and Juliet both excel at communicating their feelings about their deafness, letting us in.

I think about my great-great-aunt and the string she used to connect herself to her baby daughter. There are so many ties that connect us, generation to generation. In our family, we have passed along deafness, yes, and the particulars of a genetic code. But we pass along so much more than this: creativity and strength in the face of our genetic challenges—in the ways we frame what happens, and in the ways we cope with it all. We pass along connection. Strings, wrist to wrist.

This past year, Sophia turned 11; Juliet turned 8. One spring evening, we had guests coming over, and it dawned on Sophia that she couldn't hear the doorbell when she was upstairs in her room, even with her hearing aids on.

This bothered her. She wanted to rush down to greet our friends when they first arrived. And she wanted to know (more generally) whenever someone came to the house.

She came up with an idea.

She trailed a string from her bedroom window down to the front door. At the end of the string, she tied a basket (a small pad of paper and a pen inside), dangling in front of the door. On the top of the pad, Sophia printed a message: "Please write your name in this pad, place it back in the basket and yank on the string three times." The tugging of the string made Sophia's window blinds move up and down. When she saw the movement, she hoisted up the basket and read the name on the pad. If someone she knew was at our front door, Sophia could head downstairs to let them in.

Sophia's string.

Only for Sophia, the string led outward, connecting her not to me, or Bill, or Juliet but to the world outside. Teaching people beyond her how to reach her. Showing them that (with a bit of innovation) she could hear, and be heard.

# THIRTEEN

# What If

Laurie Strongin

..........................................................................................................

If you knew Henry would suffer and then die, would you still have had him?"

It was 6:20 one morning in the spring of 2012. I'd come downstairs to make coffee, when I saw my 10-year-old son, Joe, sitting on the couch staring at a picture of his brother Henry on his computer screen. An old picture: Henry in his ever-present Batman costume.

"Yes. Of course," I said.

Joe was only 1 year old when Henry died, at age 7, of a rare, deadly genetic disease called Fanconi anemia. In an effort to stay connected to this brother who he doesn't remember, Joe asks a lot of questions.

"Did Henry like to play with me?"

"He loved playing with you. He called you Little JoJo and loved to hold you in his lap and play peekaboo. You made him really happy."

"Did Henry know he was going to die?"

"No. Up until the end of his life, he never thought of himself as sick. He was always getting better."

"Was he scared?"

"No. He was really brave. He'd look at the nurse who was about to stick him with a needle and he'd say, 'Bring it on!' with a grin. He did a lot of hard things, but he nearly always had a smile on his face. He had big dimples, just like you and Jack."

"How did he get sick?"

"That's a really complicated question."

I explained that Henry got his disease from his dad (my husband, Allen) and me. We each had inherited a defective gene for Fanconi anemia from our mothers, which together we had passed on to Henry.

"Half from you and half from Dad? So Henry is equal to one?" Joe clarified.

He started asking about the rest of the family, beginning with himself and his 15-year-old brother, Jack. I reassured him—again—that they were both tested before they were born and neither has the disease. I explained that neither is a carrier, so regardless of who they marry, they can't have a child with the disease.

Joe was taking notes on the back of an invitation. He loves to write, to keep track. He carefully wrote "Joe" and "Jack" on the envelope, followed by the words, "No baby with diesese."

I worked my way through the family tree, explaining who is and isn't a carrier, who can and can't have a baby with Fanconi anemia. Joe wrote everything down. "Nana—carrier, Papa Sy—not carrier. Can't have any baby with diesese." "Grandma—carrier, Pop Pop Teddy—not carrier. Can't have any baby with diesese." He asked about all his aunts and uncles, carefully recording their carrier status and their potential to give birth to a baby destined to die young. By the end of the conversation, his envelope was covered with a sprawling family tree with one broken branch.

October 25, 1995. I was lying in a hospital bed, numb from the waist down, in an operating room on the maternity unit of the George Washington University Hospital, near my home in Washington, DC. Because my baby was breech and hadn't grown in my last two weeks of pregnancy, I was having a scheduled C-section. Allen and I were excited but nervous. Two weeks earlier, we'd learned that I had intrauterine growth retardation—our baby was no longer growing in utero. This isn't uncommon or necessarily indicative of a problem, aside from a potentially low birth weight, but I was being closely monitored.

"It's a boy!" exclaimed Dr. Richard Beckerman, my ob-gyn, upon delivery. Our baby weighed 5 pounds, and his Apgar scores—the measure of a newborn's physical condition—were normal. We were all relieved. One of the nurses swaddled him in a striped birthing blanket and put a little cap on his head. The only things visible were his face and hands, which peeked out from the blanket. The nurse brought him up close, where Allen and I could first lay our eyes on him.

We named him Henry.

In a split second, Allen, the doctors, and I all noticed an abnormality on Henry's right thumb. Before I could even hold him, he was whisked away and put in the neonatal intensive care unit, where sick babies go who need emergency care.

Allen rushed from my bedside to Henry's tiny crib, taking video to show me our newborn baby in intensive care, with an oxygen mask on, his tiny cries drowned out by the constant pulsing and beeping of the surrounding monitors.

Three hours later, a doctor I didn't know came into my hospital room and explained that Henry had a serious heart defect called tetralogy of Fallot. This would require open-heart surgery when he reached 12 pounds, or about 5 months old. He reassured us that this surgery had a 99 percent success rate and that Henry almost certainly would thrive once it was behind him. Hearing the words "Henry" and "open-heart surgery" in the same sentence offset the potential reassurance of "99 percent success." Up to that moment, I had no practice in dealing with hardship or serious illness. I hadn't spent a night in the hospital since the day I was born. I knew having a baby would change our lives, but I was totally unprepared to be thrust so early in motherhood into the world of serious illness.

Two weeks later, Henry was diagnosed with a rare, fatal disease called Fanconi anemia, and our lives changed forever.

Those two words snatched our innocence and naivety and swapped them for fear and the nightmare of our child's death.

Fanconi anemia sounds innocent. It kind of rolls off your tongue. Like a ride at an amusement park. *Let's go on the Fanconi anemia*. Or a dish at an Italian restaurant. *I'll have the Fanconi anemia*.

But there's nothing fun or delicious about Fanconi anemia. It is a child killer. It doesn't kill swiftly. It's a slow, insidious, painful death. And before it kills, it maims. It causes serious birth defects like tetralogy of Fallot, a blow it dealt Henry. For others, it brings anomalies such as missing thumbs, deafness, scoliosis, missing or misshapen kidneys, and shortness of stature. It may force others to live with feeding tubes as their digestive tracts are malformed.

If you survive the multiple surgeries to correct the defects wreaked upon you in utero by Fanconi anemia, the prize is nearly inevitable bone marrow failure, necessitating a bone marrow transplant. When Henry was born, the survival rate for a bone marrow transplant for the type of Fanconi anemia he had (FANCC) was appallingly grim. At that time, no one had survived a transplant without a perfectly matched sibling donor. You had to be pretty lucky to defy those odds.

In contrast, a transplant using the stem cells from an HLA-identical sibling resulted in survival rates of around 90 percent. (HLAs, also known as histocompatibility antigens, are proteins that recognize whether or not a cell is foreign in the body. Any cell possessing an individual's HLA type is recognized as belonging to that person, whereas a cell with a different HLA type is identified as an invader. HLA type is used to determine the compatibility of bone marrow, kidney, liver, pancreas, and heart for transplantation from one person to another.)

In short, unless Henry had a brother or sister who was a perfect genetic match for HLAs, he would die before he reached kindergarten.

Without really talking about it with each other, Allen and I made three quick decisions. First, we would live each day to its fullest. We would appreciate every moment we had with Henry. We would eat ice cream for dinner. Not after, *for*. We would go to Disney World not once, not twice, but over and over again.

Second, we wouldn't accept Henry's fate without a fight. We would make every call and pursue every lead until we found someone, anyone, in the medical world who would help us save Henry.

Third, we would help as many people as we could along the way. We felt lucky to live in Washington, DC, where we had access to great medical care. We both had the benefit of college educations and, through our families, financial resources to devote to reversing Henry's fate. We both had backgrounds in marketing and communications we could use to raise money for research into Fanconi anemia, to increase the number of people on the national bone marrow registry, and educate people about FA and genetic disease.

It was hard to accept the reality that Allen and I had mixed a deadly genetic cocktail just by falling in love and starting a family What if there had been screening and we'd known much earlier that we were both carriers of Fanconi anemia? Would we have gotten married anyway? Would we still have had babies, hoping they'd be born healthy? Would we have used donor sperm or donor eggs to alter our unlucky gene pool? There are other "what ifs."

I don't know the answers.

Allen and I didn't have to wrangle over these issues because, before we got married, we both underwent carrier screening for all known Jewish genetic diseases. We both tested negative and put aside any thought of passing anything deadly along to our children. This first time, with Henry, our decision to have a baby was uncomplicated. We loved each other. We were committed. We wanted to build a family together.

The second time was different. Now we knew we had a 25 percent chance that our baby would be born with a fatal disease. But there was a 50 percent chance the baby would inherit only one copy of the gene, making him or her what is called a "healthy carrier." And there was a 25 percent chance the baby could avoid a copy of either affected gene. We knew the probability of good health was on our side. We also knew that if we stopped having children, Henry would die.

Despite our shaken confidence, the decision was easy. We would go ahead and roll the dice and pray like hell that this time around we'd get lucky.

In May 1996, not long after Henry had turned 6 months old and recovered from successful open-heart surgery, we learned that I was pregnant again. Like any couple trying to have a baby, we were euphoric. But fear and uncertainty over the baby's health quickly crept in. Before we could figure out who to call first—our family or our doctors—the phone rang.

It was Dr. Arleen Auerbach, a preeminent Fanconi researcher. She and I spoke for a long time, and when I finally hung up the phone, I felt stunned. And elated.

I asked Allen to come downstairs, and I posed the very question that Dr. Auerbach had just asked me. "What would you say if I told you that we could get pregnant and know that our baby was going to be healthy?"

"Well, isn't it a little late for that?" asked Allen.

"Seriously, just tell me. What would you say?"

"I'd say sign us up," he said, confirming the obvious.

"What if we could also know that the baby was a bone marrow match for Henry?" I added.

"I'd say we'd found the golden ticket," he replied.

"Well, I think we just did. It's called PGD."

PGD, or preimplantation genetic diagnosis, was a cutting-edge, newly available process that could guarantee that our next baby would be both healthy and HLA-matched to Henry. Dr. Mark Hughes—chief of the Section on Reproductive and Prenatal Genetics in the Diagnostic Development Branch of the National Center for Human Genome Research at the NIH and a pioneer in the field of reproductive genetics, specifically, in single-cell genetic analysis—had figured out a way to combine in vitro fertilization with genetic testing conducted prior to embryo transfer. PGD would enable us to identify and implant an artificially conceived embryo that not only was healthy but could also be Henry's savior. By collecting this healthy baby's umbilical cord blood at birth and transplanting the stem cells to Henry, our baby—Henry's brother or sister—could save Henry's life.

Dr. Hughes had used PGD in the past to screen embryos for fatal childhood diseases such as Tay-Sachs and cystic fibrosis, enabling parents to know at the outset that their babies would not be born with a fatal disease. But neither he nor anyone else in the world had ever used PGD to identify a perfect HLA match, from whom umbilical cord stem cells could be harvested and thus save a sibling.

"We'd be the first?" Allen asked me.

"Apparently so."

PGD involves the biopsy of one or two cells from an eight-cell embryo, typically on the third day following egg retrieval, as part of an IVF cycle. The biopsy is performed in a laboratory by making an opening in the outer "shell" of the embryo and extracting the cell(s). After the biopsy, it takes about forty-eight hours for genetic testing to be completed before the embryo, which remains in a lab and continues to develop to the blastocyst stage, must be transferred to the woman's uterus to produce a viable pregnancy.

The extracted cell or cells are analyzed to determine the genetic composition of the embryo. These tests can determine the presence of chromosomal abnormalities such as Down syndrome. Testing can also be done for couples known to carry fatal diseases caused by a single gene abnormality, like FA. The test results are used to inform the selection of embryos for transfer to the woman's uterus, enabling her to begin her pregnancy confident that her baby will be healthy.

Physicians like Dr. Hughes have used PGD for prevention of fatal disease for more than a decade. More than a thousand babies have been born healthy through this procedure. But in 1996, when we began talking with Drs. Auerbach and Hughes, PGD had *never* been used to test for HLA type or any other trait that involved something more than the survival of the embryo itself.

For the first time ever, Dr. Hughes would offer PGD with HLA typing to families with a history of Fanconi anemia who met very specific criteria. First of all, the mother had to be under 35 years old. (Younger mothers typically produce more eggs in the process of IVF and therefore have a greater likelihood of success. They also have a lower risk of producing eggs with abnormal chromosomes.) In addition, Dr. Hughes would work only with families who genuinely wanted additional children. Babies born through this technology must be wanted, independent of their ability to save a sibling's life. Lastly, the family had to have the FANCC, IVS4 Fanconi mutation, because this was the only type of FA for which a gene had been discovered, meaning it was the only type that PGD could diagnose.

Only two families met these criteria. Allen, Henry, and I were one.

If we pursued PGD, Allen and I would be entering uncharted territory. As the first people in the world to use this technology for this purpose, there was no precedent to consult, no books or articles to read, no support groups. There were neither established ethical guidelines nor government or industry regulation of PGD. There would be no guarantees.

First, we'd wait for the results of prenatal testing that would determine whether or not the baby I was already carrying was healthy. If so, regardless of his or her HLA type, we would continue the pregnancy. As first-time parents of a baby newly recovered from open-heart surgery, with inconceivable hardship waiting for him on the horizon, we didn't have the luxury of time, energy, or imagination to figure what we would do if our second baby had FA. But we were grateful to know that we had choices.

PGD was our lifeline out of this horror.

Allen and I carefully considered the ethical implications of PGD. We had so many things to consider, and so little information. We were determined to protect and advocate for Henry, and for our future children, while also honoring our values and those of the broader community.

For more than half a century, prenatal diagnosis through procedures like amniocentesis and chorionic villus sampling has enabled couples to screen fetuses for genetic disorders. Since that time, couples like us have faced the difficult dilemma of whether to proceed with a pregnancy, based on these test results. The fact that PGD screens embryos before implantation, and not fetuses already growing in the womb, held tremendous appeal. Allen and I are both pro choice, but we did not want to have an abortion, particularly to avoid having another child like Henry.

We also understood that PGD would result in the creation of embryos that would not be selected for implantation. This is true for almost any couple undergoing IVF, in which robust, healthy-appearing embryos are implanted, and those failing to develop are not. People who regard embryos as human beings object to creating and discarding embryos and, consequently, are opposed to IVF and PGD. But IVF is legal and widely available and is pursued by tens of thousands of couples each year. As a result, thousands of embryos are implanted, while hundreds of thousands are cryopreserved for future use, donated for medical research, or discarded. Our embryos would face a similar fate. We believe embryos deserve special respect, as they possess the potential to develop into humans once they are implanted into a uterus, but we do not view them as people. This cluster of undifferentiated cells does not look like a human being. It cannot reason, and it cannot survive for even a moment outside the womb. In contrast, Henry was alive. He could think, feel, smile, and laugh. We felt ready to go ahead with this. Because we wanted additional children, we intended to freeze excess embryos for future use.

Testing for HLA—a trait critical to the success of Henry's stem cell transplant, but not to the survival of the potential child we were testing—was brand new. HLA type had never been tested in a lab prior to pregnancy as part of a stem cell donor search. As such, it had yet to undergo extensive ethical review and so had not earned the status of moral acceptability. Together with our doctors, we would have to navigate this yet-to-be-explored medical frontier.

Years later, when news of the use of PGD for disease avoidance and HLA typing first broke, it was often grouped with selection for parental preferences such as family balancing for a child of a certain sex, sports ability, or IQ. It got reduced to the politically charged term "designer baby." I have a hard time understanding why anyone would go through the grueling physical agony and phenomenal expense of PGD solely to produce a daughter to help complete the "perfect" family (mom, dad, son, daughter) or to have a child of Olympic caliber (not that it is even possible to test for that). We strongly believe that the selection of an embryo based on its HLA type does not belong in that conversation. Our baby's HLA type was not a superficial desire but a matter of life or death.

Others rightly raised objections to the use of a baby's organs as spare parts. Removing a young child's kidney or liver to aid an ailing sibling is risky and disturbing and merits serious debate. But using a baby's umbilical cord blood—which was all we planned—is different. The blood in the umbilical cord, where the stem cells used for bone marrow transplants are found, is typically disposed of as medical waste after delivery. Retrieving these cells causes no harm to the baby. Essentially, it is life-saving recycling.

Allen and I gave these issues a lot of thought and discussed them at length with our doctors, but we had no hesitation about moving forward. We believed that if we worked with brilliant and compassionate physicians using revolutionary technology, we could reverse Henry's fate. It was our best, and likely our only chance to save him.

Two months later, prenatal testing revealed that the baby I was carrying was healthy. While the first time around I had taken good health for granted, this time it felt like a miracle. Knowing PGD was available to us down the road, the news that the baby was not an HLA match to Henry seemed of little significance.

We placed our faith squarely in the promise of PGD.

Allen and I began to prepare for our first PGD attempt while I was still pregnant with Jack, our second son. We provided blood samples for DNA analysis and HLA typing, met with our doctors, and educated ourselves as best as we could. We would start in earnest shortly after the new baby was born.

PGD begins with IVF, which meant that getting pregnant for the third time was more complicated, painful, and impersonal than my two previous pregnancies. When Jack was a few months old, I would travel to

New York City, to the Weill Medical College of Cornell University's Center for Reproductive Medicine and Infertility, where I would begin the IVF process with daily injections of a drug called Lupron to halt my natural egg production. Weeks later, I'd add an additional injection of hormones for six to eleven days, to stimulate superovulation and produce, we hoped, twenty or more eggs each month. When the eggs matured, I would take a single injection of a hormone called human chorionic gonadotropin to trigger their final ripening. Within the next thirty-six hours, doctors would surgically retrieve my eggs and fertilize them with Allen's sperm. Three days later, one or two cells would be extracted from each embryo and sent to a reproductive genetics lab, to determine which embryos were both FA free and an HLA match to Henry. Within two days, the embryos meeting those criteria would be transferred to my uterus. After the procedure, I would take daily injections of progesterone until my pregnancy test. All told, this procedure would take about five weeks. Nine months later, we would have a healthy baby—our third child—and be assured that Henry would have the life we'd meant to give him at birth.

It sounded too good to be true.

When we began researching PGD in 1996, it was neither on the national news nor featured in fertility clinic advertisements. At that point, it was somewhere between a distant hope and an extraordinary dream shared by a small group of doctors and families who believed in the promise of science.

On January 9, 1997, when Jack was just 13 days old and we were weeks away from starting PGD, our dream of saving Henry turned to a nightmare. There in the morning's *Washington Post* I read that Dr. Hughes's work on PGD had been accused of violating the federal ban on embryo research, known as the Dickey Amendment. Although Dr. Hughes did not conduct any embryo research in his position at the NIH, he was charged with violating the Congressional ban because his PGD work at nearby Suburban Hospital—where he was already working on our case—employed NIH research fellows and scientific equipment, specifically, a refrigerator, that the NIH said had been moved to Suburban without NIH approval. While I read the words on the page, the only thing I could see, as if it were written in black and white before me, was one sentence.

*Henry is going to die.*

According to NIH sources cited in a *Chicago Tribune* article that same day, Dr. Hughes got into trouble because one of the four research fellows assigned to assist him, all funded by the NIH, was worried that experiments they were conducting violated federal law. Apparently, the fellow reported his concern to Genome Center officials, who ordered the investigation that resulted in Dr. Hughes's termination. I'm *not* a policy expert, nor am I supportive of the federal ban on embryo research, but it is inconceivable to me that a refrigerator and one research fellow separated a brilliant doctor from the only chance that my son might live beyond the age of 5.

Within weeks, Dr. Hughes lost his job and faced Congressional hearings that vilified him and threatened both his career and our prospects. Months passed, we waited, and, as predicted, Henry's blood counts fell. Because no other doctor in the country was engaged in groundbreaking PGD research, there was no one else to turn to.

Dr. Hughes refused to allow politics to destroy his science, and seven months after the *Post* article first appeared, he set up shop at a private lab at Wayne State University in Michigan, where he could resume his life-saving work.

Between December 1997 and June 2000, I tried PGD nine times. During that period I received 353 injections, produced 198 eggs, yet had no successful pregnancy. Often our most robust embryos had Fanconi anemia, while the poorest quality were Fanconi-free HLA matches that failed to produce a pregnancy. Allen and I spent nearly $135,000, most of which was not covered by insurance, and spent far too many days apart from one another, our two sons, and our lives. Our hopes were raised to the highest heights and crashed to the depths of despair, over and over again. There was no medical explanation for our lack of success, just bad luck. It had never occurred to me that people on the frontlines of new medical discoveries are rarely the beneficiaries of their promise. Now, I was learning that the hard way.

Life went on during this time. Henry became a little boy. He fell in love with a girl named Bella. Donning a blue blazer and khaki pants, he dutifully attended her ballet recital and presented her with white roses after her performance. Jack quickly became Robin to Henry's Batman, and our two caped crusaders befriended all our neighbors as well as the shop-keepers in our close-knit community. Henry displayed acumen on the soccer field, scoring goal after goal in each game, and formed lifelong bonds with his family and friends from preschool. That was one part of the story. But at the same time, Henry's platelets fell from a high of 103,000 to a low of 10,000; his hemoglobin, from a high of 12.2 to a low of 6.9; his *absolute* neutrophil count, from a high of 1,900 to a low of 300. By age 4½, Henry's bone marrow was failing.

We had run out of time.

We faced a choice we'd spent years evading. Would we pursue a bone marrow transplant with an unrelated (non-sibling) donor, knowing the survival rate was alarmingly low? Although the success rate was no longer 0 percent, as it had been when Henry was born, at approximately 28 percent, it didn't provide much comfort, particularly given that the stakes were our son's life. But if we refused the transplant, Henry had 100 percent chance of dying.

In July 2000, Allen, Henry, Jack, and I drove from our home in Washington, DC, to Minneapolis. Henry's transplant procedure would take place at the University of Minnesota Children's Hospital, with a donor identified through the National Marrow Donor Program. This was the very same hospital where the first pediatric bone marrow transplant had been successfully performed. It featured some of the world's preeminent Fanconi specialists, who had performed dozens of transplants. We knew that this place and these doctors gave us the best possible chance that Henry could be among the first patients with his particular type of FA to survive an unrelated bone marrow transplant. Again, we would be pioneers, hoping this time that the medical breakthrough would help us.

Initially, Henry fared well, despite the havoc instigated by the chemotherapy, radiation, and foreign stem cells transplanted into his body. After a few months, though, he started to suffer serious side effects that persisted despite the seventeen daily intravenous medications he received and dozens of surgeries he underwent during hundreds of days and nights in hospitals from Washington, DC, to Baltimore, MD, Hackensack, NJ, and Minneapolis, MN.

Throughout that time, Allen and I never gave up on our desire to have a third child. In February 2001, we tried to use some of our frozen embryos produced during our pursuit of PGD. Many of the embryos were known to be Fanconi-free, but because they weren't HLA-matched to Henry, we didn't use them the first time around. Despite defrosting nearly twenty embryos, only two of which were vibrant enough to implant, I didn't get pregnant. Not long afterward, Allen and I signed consent forms donating our nearly one hundred remaining embryos to Dr. Hughes for research that would provide hope and answers to other couples, so no one else would ever have to live through this heartbreaking devastation.

Finally, we opted for the same Russian roulette–style method of conception we'd used to have Jack and, mercifully, got pregnant with a third boy whose health was confirmed through chorionic villus sampling in my first trimester.

On October 6, 2001, I gave birth to our third son, Joe. In all the complexity of our lives, Joe came to us easily. We'd tried to produce him through a mix of scientific breakthrough and miracle, but instead he showed us that sometimes the best things in life come simply and naturally.

Despite the beating that Fanconi anemia dealt Henry on a nearly daily basis for the two-and-a-half years following his transplant, he was a happy kid. He charmed his doctor into letting him out of the hospital with a two-hour pass so he could attend his friend's birthday party. He wooed a nurse into taking him on a date to his favorite restaurant, Cactus Cantina. He went to the movies if he felt well, or hosted movie parties in his hospital room if he didn't. He attended kindergarten, achieving the heartbreakingly simple yet nearly impossible goal we set when he was just 2 weeks old. He collected Pokémon figures, foreign coins, and marbles. He met President Clinton and his hero, Baseball Hall-of-Famer Cal Ripken. He rode on the Batcycle and in a life highlight, met the real Batman. He was a great brother to Jack and Joe, and an incredible son. Henry knew how to live and taught everyone around him how to do the same.

Being with Henry made everything better.

Which made his death devastating.

On December 11, 2002, at 6:40 p.m. CST, Henry died in my arms. The official cause of death was aspergillus, an untreatable fungal infection in Henry's lung. But it was really the failure of PGD that killed my son.

Shortly after Henry died, I was talking with a friend of mine who, years earlier, had suffered a similar loss. I asked him how he managed to live without his daughter. He explained that he didn't have to. He pointed to his shoulder and said, "She's right here with me. I carry her with me wherever I go." Drenched in sorrow, I didn't know how in the world Henry could come back to me.

As we approach the tenth anniversary of Henry's death, now I understand what my friend was talking about. Henry lives on in the hearts and minds of everyone who loved him. He lives on through the children whose lives have been saved through our pioneering work on PGD. Henry shares his ability to live well and laugh hard even in the face of serious illness every time the Hope for Henry Foundation—the foundation Allen and I started in 2003, to honor him—gives a kid with cancer or blood disease a new iPad, or the opportunity to meet Batman or an acclaimed author, or to go to a summer carnival or Halloween party, or have a birthday party while in the hospital. That's 6,500 kids and counting. Henry did more in his seven years than most octogenarians do in their lifetime. Which is why, without hesitation, I am glad we had him, even though he suffered plenty and died young.

# FOURTEEN

# The Long Arm

Clare Dunsford

·················································································

As mutations go, the one my family carries is particularly per-
verse. Part of a group of mutations described as "expanding" or
"dynamic," because the gene changes in size over generations, fragile
X—our family curse—cannot be explained by Mendelian inheritance, for
the gene is neither completely dominant nor fully recessive. Initially, the
pattern of inheritance of fragile X was so contrary to what scientists had
known until its discovery that, when it was understood, it was dubbed
the "Sherman paradox," after the scientist who explained its baffling
characteristics in 1985, the year my son was born.

What puzzled doctors was that if a condition is X-linked, it usually
affects only males, but fragile X affects females, too. Moreover, the
males who passed on the fragile X mutation to their daughters seemed
unaffected by it. What Dr. Stephanie Sherman figured out was that the
seemingly unaffected mothers and grandfathers of boys with fragile X
were actually carrying a premutation of the gene. In other words, they
were not genetically normal, although they were phenotypically so.[1]

The FMR1 gene is located on the X chromosome, on what is called
the long arm (all chromosomes have a short arm and a long arm, pinched
in the middle by the centromere). Fragile X syndrome is one of a group
of conditions called trinucleotide repeat disorders, in which a section of
the gene increases in size, duplicating itself in a disastrous stutter to the
point that the gene becomes unstable. A typical individual has fewer than
45 repeats of a phrase that reads CGG on the FMR1 gene, but someone,
like me, with the premutation of the FMR1 gene has between 55 and
200 repeats. (There is a gray zone of 45 to 55 repeats in which the gene

may or may not expand.) When the premutation is passed on, it expands further, and once the repeats spin off beyond 200, the gene shuts down production of its protein. The threshold of the "normal" world is reached, and the unlucky person with the full syndrome of fragile X is born with cognitive disability, severe anxiety, and attention deficit, and often with seizures.[2]

To add to the perversity of the fragile X mutation, males who have fragile X syndrome and all that it entails cognitively, emotionally, and physically will father normal offspring. That is because their Y chromosome is normal and thus can produce only normal boys, and their X chromosome, ironically, passes on only the premutation of the gene, so all their daughters will be carriers, like me. To put this in personal terms, my son, whose IQ is 54 and who cannot make change or drive a car or live independently, would have a normal child, in the unlikely event that he should reproduce, but I, the woman writing this essay, have only a 50-50 chance of doing so. If my child receives my mutated X, the odds are almost 100 percent that my premutation will expand into a full mutation. Talk about a paradox.

When my son was diagnosed with fragile X syndrome in 1992, the gene itself had been discovered only the previous year, and it was only slowly that I grasped the odd way in which I had passed it on to him. When I was told my son's failure to develop normally was due to a genetic disorder, my first thought was that there was no history of mental retardation in our family (what was this doctor talking about?), but as time went on, I understood that unlike most inherited disability, fragile X didn't leave a visible trail. It was a silent and slowly growing mutation that geneticists call, with no intended irony, a mutation of anticipation. Rather than offering a hope for the future, this mutation seemed to bring my family to a dead end in the present. Genetically speaking, however, my son's problems were indeed "anticipated," albeit in an insentient way, by the changes in my father's and then his daughters' FMR1 genes. The mutation was pointed like an arrow into the future—poised to strike our children.

What I didn't know, but have come to find out, is that the arrow is actually a boomerang, and my sisters and I, as well as our father, are right in its path.

# The Proband

On a recent visit to the gym, I overheard one woman say to another, "My youngest is about to graduate from high school." "Welcome to the empty nesters!" the other replied cheerily as she pulled down the lat bar. I simply continued my bicycle twists on the mat, hoping they wouldn't include me in the conversation. My son, J.P., is now 27, but my nest may never be empty. He can't live on his own, because fragile X syndrome has robbed him of normal intelligence and the ability to carry out daily tasks without a lot of supervision.

Even before genetic inheritance was understood, humans recognized that, as often as not, like begets like. So it was that when I became pregnant, I pictured my child with brown hair and eyes, like his father, Harry, and me; I assumed that he or she would do well in school like we did, and laughed to think that it was 50-50 whether he or she would be athletic (Harry was, I was not). That is who we "anticipated." But it is not who we got.

Blond and green-eyed, our baby boy still had his father's strong brow and so bore the imprint of his father's genes. Yet from birth, J.P. was floppy in my arms, a sign of low muscle tone that was a red flag for pediatricians, indicating that developmentally, something was wrong. His pediatrician also noticed that J.P. didn't make eye contact; his eyes darted off to the side, a condition called nystagmus. She ordered a CT scan of his brain, fearing a tumor, but what the scan showed instead was that the ventricles of J.P.'s brain were mildly enlarged, and he had wide extracerebral spaces that suggested megalocephaly. The neurologist we met with explained that the doctors really didn't know the significance of these findings, but what they had discovered was clearly abnormal.

As the months went on, I caught and tried to deny subtle signs that J.P. wasn't developing as most babies did. While Harry's mother bragged of his early walking, and my mother of my early talking, our own baby crawled late, walked late, and talked late; he was obviously not taking after us. Meanwhile, well-meaning relatives and friends told us not to worry, that Einstein didn't speak until age 4, that I was just a nervous first-time mother. But while I was always a nervous Nellie, I wasn't stupid. I could see that J.P. didn't play like the other toddlers in the play group I joined (and left quickly). Eventually, Harry and I sought out an expert opinion, and so began our pilgrimage from one Boston neurologist to another, seeking to understand better why J.P. acted the way he did.

Subsequent neurologists we consulted diagnosed J.P. severally as having "language delay and attention deficit"; extreme ADHD, attention deficit and hyperactivity disorder; and most seriously, PDD, or pervasive developmental disorder. Finally, a neurologist at Boston's Franciscan Children's Hospital raised the possibility that J.P. might have a genetic disorder, fragile X syndrome. That name was new to us.

When a blood test in late 1992 confirmed that J.P. did indeed have fragile X, he was 7 years old. By now Harry and I were divorced, and my life was already altered beyond all expectation. A few months later, when I met with a geneticist, I began to learn the ramifications of this new diagnosis. This "dynamic mutation" proved to be sheer dynamite in the Dunsford family. When I made the phone call to my parents to tell them I'd learned J.P.'s condition was genetic, I could never have imagined that the bombshell I dropped would continue to fling shrapnel twenty years later.

I am the oldest of five siblings—four girls and a boy—and my son was the first grandchild. At the time of my phone call, one sister already had two children, another had one girl (and would give birth to a second a few years later), a third was engaged, and my brother's wife was pregnant with their first child. The stage was set for a cascade of consequences. J.P. is what is called a proband, the first of a family to be identified as having a genetic disease, the canary in the coal mine. A genetic diagnosis requires a family to range backward and sideways to discover whether anyone else carries the gene. All my siblings had a stake in what I'd found out, for the geneticist told me that even if my nieces and nephew seemed normal, every one of us needed a DNA test to see whether we carried the premutation that results in fragile X.

The results came back almost (but not quite) as bad as they could be. All of my sisters carried a premutation of the FMR1 gene; only my brother's X chromosome was normal. That meant my sisters' three children had to be tested, as well—two girls and a boy—and it turned out that both girls, one belonging to each of two sisters, tested positive for fragile X syndrome. J.P. and I were now not alone, but my pain was tripled knowing that my sisters shared my genetic fate. Maggi's daughter and Ann's daughter had their father's normal X as well as the mutated X, and so they weren't as severely affected by the mutation as J.P. This is typical of females with fragile X, but as the years went on, their symptoms became more burdensome: seizures, anxiety, shyness, learning disabilities, depression.

When family members share a genetic disorder, it means that when I tell my story, I necessarily bump up against theirs. One of my biggest challenges as a writer has been to respect the privacy of my sisters and my nieces who, after all, have their own stories. So it is J.P., the proband of this tale, who stars in the story I tell. John Patrick, my first and only, the child I never anticipated and the child I couldn't imagine not being mine. Now a young man, he remains my child in ways peculiar to those who develop differently than typical children. He is a living paradox, a man who kisses his "mama" goodnight and sleeps with a worn-out Winnie the Pooh and a tattered stuffed cow named Isabelle, but who can now work in a college library four hours a week with the aid of a job coach, where he notes that the B.C. girls are "hot chicks."

Anxiety is one of the hallmarks of fragile X, in carriers as well as in those with the full mutation. While low IQ bars the door to many developmental experiences for those like my son—understanding math and time, driving a car, marrying and creating a family—anxiety slams that door shut. New experiences of any kind, whether simply entering an unfamiliar store or shifting a daily routine, flood the body of those with fragile X with cortisol, the hormone that rises with stress. When J.P. was younger, fear or anxiety would make him erupt in catastrophic outbursts, sometimes hitting those around him or biting himself or others, like a wild animal in a trap.

I could never have imagined during those challenging early years that J.P. would function as well as he does today. In the past several years, J.P. has mellowed to the point that he's eager to try new experiences; he can work a party like a seasoned politician. I'm tempted to believe that we are seeing the effects of minocycline, which J.P. started taking in 2008, in addition to two other medications he'd been on for a while. Minocycline is a common, well-established antibiotic that is being used experimentally by some in the fragile X community to lower anxiety and potentially build new connections in the brain. Studies of rats whose FMR1 gene has been knocked out, models of my son's neurobiology in important respects, have demonstrated an increase in tolerance of novel experiences and greater cognition when given this drug.[3]

Hearing the nightly news on TV recently, J.P. asked, "Where is Syria?" a simple question he would never have asked a year ago. "What's that mean?" he'll say as he hears me talk about a student taking an "Incomplete" in a course at my university. Curiosity is a luxury when your brain buzzes with inchoate sensations, but it's one that he can now indulge.

I'll never forget what happened when J.P. had taken minocycline for only about ten days. It was a new thing for me at that time to leave him home alone for short periods, and it was still exhilarating and nerve-wracking. I came home from a short errand and saw the house key that I kept hidden for emergencies lying on the counter. I had a sudden implausible thought. "J.P., did you go out?" "Yep," he said casually, as if he regularly went on expeditions of his own. (A true couch potato, my son usually has to be dragged outdoors, and spends most of his time in front of the TV or the computer.) "Where did you go?" I asked incredulously. "On a walk," he said. "Where?" "Around the block," he replied nonchalantly. Ever the mother, I asked, "Did you wear a coat?" "No," he said, the "duh" all but audible.

While this incident sounds unremarkable, it was unprecedented for my son, and it filled me with fear and hope: fear that he might get hit by a car, to name one hazard, and hope that he might become more independent. Fear and hope dominate my horizon these days, as a plethora of clinical trials of new drugs have flooded the fragile X community. J.P.'s father and I must make the difficult decision for him whether to take on the risks that new compounds pose or be satisfied with the impressive progress he's made so far. Drugs whose names sound like Star Wars characters—STX209, AFQo56—are so-called mGluR5 antagonists, which target the out-of-control glutamate signaling in the brain that causes the symptoms of fragile X.[4] The boy we anticipated, the boy we got, the boy we might have—we love them all, but what a wild ride we are on along the genetic frontier. All parents watch as their child matures and changes, but for us, change is jump-started by a chemical fresh from the laboratory.

Among the unpleasant things you get used to when you receive a genetic diagnosis is the experience of hearing yourself or your loved ones described in clinical language that makes you sound like an alien, a primitive, or an animal. Genetic vocabulary sometimes rivals the worst insults.

So it was when I first heard the grandfathers of children with fragile X referred to as *nonpenetrant males*. To me, that phrase conjured up an image of an emasculated and inconsequential fellow, a kind of Dagwood of the genetic world. Literally, it means that a trait is present in the genetic makeup of a man but does not express itself in him or, in the case of fragile X, in his offspring. The trait only penetrates the third generation, the grandsons, and there it flowers into a full-blown syndrome.

Since, of my parents' offspring, only the girls carried the premutation and my brother was free of it, it seemed likely that we inherited it from my father, who would have passed on his Y chromosome to his son and the X to his daughters. My father, the "nonpenetrant male," is a strong and successful man, now 85. He endured the loss of his own father when he was 9, almost died from acute appendicitis when he was 12, excelled in high school, college, and law school, edited his college newspaper, fought in the Korean War, fathered seven children, lived through kidney cancer and two bouts with colon cancer, held a chair at St. Louis University Law School, became a nationally known arbitrator, and became the president of the elite National Academy of Arbitrators. My father is intelligent and articulate, a superb writer, a dynamic teacher and lecturer, a passionate Catholic, and fiercely upright.

But in the pedigree the geneticists drew of our family—another offputting scientific term that reduces family history to a dog's bloodline—this talented man was deemed a "nonpenetrant male." He it was whose FMR1 gene had first gone rogue, but it had not visibly altered his path in the world. At least not yet.

I was at one of the biennial conferences of the National Fragile X Foundation in the early 2000s, and at one of the sessions, Dr. Randi Hagerman—the leading clinician of fragile X, a developmental pediatrician and researcher—asked an audience of carrier mothers whether any of their fathers were showing certain symptoms. "How many of you have fathers who have balance problems when they walk?" Several hands were raised. "How many of your fathers have a tremor?" More hands. "How many of your fathers are having problems with anxiety?" More hands. It was the dawn of a new age in the world of fragile X. Randi and her husband, Dr. Paul Hagerman, a molecular geneticist, had discovered that the grandfathers of the kids Randi saw exhibited symptoms of a syndrome of their own. In time, they dubbed it FXTAS, fragile X tremor-ataxia syndrome.[5]

I didn't raise my hand at that conference session. My father didn't seem to show any of the symptoms Randi Hagerman described. Yes, he was a worrier and pretty high-strung. That's where I got it, my mother would say before we ever talked about genetics, back when I was a little girl. But he seemed healthy otherwise.

But as the decade went on, my father suffered increasingly from neuropathy in his legs and began to walk with a wide, unsteady gait. He used to like to walk around the park across from our family home; it was a mile and a quarter around and attracted scores of runners and walkers at all times of the day. My father, like his mother, like me, walked fast, pressing a long stride into the sultry St. Louis air or the crisp days of fall. It felt good to move his limbs after the sedentary work of a professor. Truth be told, though, my father never sat still for long; his nervous energy kept him jumping up from his desk at short intervals. If he heard the back door open at home, he was out of his home office in a flash to see who it was. As a young professor, his office was on the top floor of the law school, and he never took the elevator. He never even simply walked the stairs. He vaulted up the six sets of stairs, two steps at a time. I myself often take stairs that way, imprinted as a child with my father's exuberant energy.

Now, that energetic gait was slowed. Then a tremor set in when Dad tried to open a jar of mustard or write longhand. He fumbled items that he held, dropping them more often than he used to. His short-term memory began to fail. He was getting old, he said, and these things were to be expected. Yet his mother, who lived till 88, was spry and agile until her final illness with colon cancer. Even in her early eighties, she got up on her second-story roof to clean the gutters.

It was more than age that was settling in. It was, though we denied it for years, a blasted gene that was always there, in every cell of his body, an evil intimate that decided to take down my father just as it had taken down my son. I guess it had finally penetrated, like the cold in an uninsulated house.

Why is it that the disability caused by a genetic mutation seems more ominous, more oppressive, than the disability caused by simple aging? Perhaps it doesn't seem so to the sufferer, to my father, but it does to me. The source of our pain matters; the pain of labor is bearable in a way that the pain of bone cancer is not, but the pain of cancer may be borne more easily than that of deliberately administered torture. For better or worse, human beings tend to make meaning out of suffering. Although, as my father points out, many of the symptoms of FXTAS are similar to those of simple aging, FXTAS is aging on steroids. Every ailment that old age can (but does not always) bring is bundled into one paralyzing mess.

When it comes to affliction, the backstory matters. My father had already seen his children suffer and his grandchildren's potential impaired. To add a degenerative neurological condition to that adds existential insult to injury. My father's diagnosis carries with it the weight of Greek tragedy, a fate set in motion by a spiteful double helix that marks the spot of origin with an X. No matter how he lived his life, Dad's premutation was destined to rob his last years of ease and dignity.

Yet I guess I should say my father is "lucky" not to have developed symptoms of FXTAS until late in life, as some grandfathers of children with fragile X exhibit them even in their sixties, cursing these men with a much longer slide down than a man in his eighties. There is no cure for FXTAS yet, any more than there is for fragile X syndrome. Drug treatments only ameliorate the symptoms slightly—a patch on a leaking tire that eventually forces the car to the side of the road, leaving it marooned and helpless.

## Obligate Carrier Daughters

The clinical term for the carriers of a fragile X premutation, the genetic term *obligate carrier daughters*, has a Biblical ring to it and conjures up, for me, a line of Egyptian slave women, each with an amphora of water on her head, walking single file on a dusty road. It describes my three sisters and me in the genetic pyramid our family inhabits. It's hard to say whether the apex of the pyramid is J.P., the proband, or my father, the first generation to mutate, but all I know is that we women are in the middle, bearing the load. In biological terms, the role we play was foisted upon us, inevitable once Dad's FMR1 gene started to stretch out. We were carried into our destiny as much as we later carried the germ of our children's fate. At this point, in our forties and fifties, my sisters and I have dim memories of the country from which we came, its carefree games and unexamined expectations, but for the most part we have adopted the strange customs of our new land: blood draws, the scrutiny of the microscope, a blighted identity marked by new names (carriers, mutants). We trod on, tied together by ropes, not of hemp, but of twisting DNA.

So-called obligate carriers were long thought to carry a mutation but to remain healthy themselves. As obligates, they were absolutely essential to the genetic story line: their children could not develop the full mutation from their grandfathers unless the carrier daughters saw the story through. In fact, I thought that I became a "carrier" only when I gave birth to a child with the full mutation, since otherwise, what I carried would never have manifested itself. My new identity was not only contingent but temporary, I thought, receding into the past as my child grew older.

As genetics has advanced, however, it has become apparent that premutation carriers are not pure vessels of delivery, a means to an end. We are affected by what we carry; its weight strains our muscles, the water sloshes down over our heads. The first pathology identified in fragile X carriers was a higher incidence of a form of premature menopause (primary ovarian insufficiency, or POI, in medical terms) in about 20 to 25 percent of carriers.[6] Then researchers detected higher rates of depression and anxiety among my cohort, although given the prevalence of these psychiatric disorders in the general female population, I'm not entirely convinced.[7]

But then came FXTAS. At first, we were told that only the grandfathers developed the disease. Then we heard stories of a mother here and there who had some of the symptoms. Now both the scientific and the anecdotal grapevines are abuzz with more and more reports of women who have a tremor or who have balance problems. The most recent conference of the National Fragile X Foundation featured a lunchtime workshop called "It's Not About My Kids: It's About Me." That pretty much sums it up. Just as I was beginning to bask in J.P.'s progress and harbor hopes for his future that had seemed unimaginable when he was a child, I now have to worry about my sisters' and my own health. Over the past few years, I tried to stop my ears to what I heard at the foundation's conferences. "I can't go there!" I told my sister Maggi, and she agreed. "Taking care of our kids and worrying about Dad is all we can handle."

The other shoe fell in late 2007 or early 2008. I was on the phone from my home in Boston with my sister Cathy, who like the rest of my family lives in St. Louis. "Are you OK?" I asked. "Your voice sounds funny." "Oh, yeah, my tongue feels swollen. It's been feeling kind of weird [it sounded like 'weahd']." Her speech was subtly blurred. I felt a cold chill, since I had known two women whose first symptom of ALS, Lou Gehrig's disease, was an unruly tongue. I urged Cathy to see a doctor, but as her speech difficulties came and went over several months, she brushed them off.

The slight slur to Cathy's speech eventually became pronounced; her vowels were exaggerated, and her words stretched out as she tried to control her mouth to form them. I was shocked when I next visited St. Louis and heard her garbled speech. She had been doing telemarketing from home, and one day the person she'd called had accused her of being drunk. Soon she couldn't control her tongue at all, and then one day she couldn't even close her mouth. She had to turn her head upside down to try to wrestle her mouth into compliance.

Cathy finally saw a neurologist, who diagnosed her with an oromandibular dystonia. The word *dystonia* was not unfamiliar to my family, as one of my nieces with fragile X suffers a dystonia of the neck called torticollis. My sisters and I tried, and failed, to repress the suspicion that Cathy's condition was related to being a carrier of fragile X. Finally, Cathy was seen at the MIND Institute at the University of California, Davis, by Dr. Randi Hagerman and a flock of other researchers, who tested her every way they could imagine.

The day before Thanksgiving 2008, my sister Maggi called me with the awful news that Dr. Hagerman had concluded that Cathy had a form of FXTAS they had not seen before. My anguish and my anger that day were unlike any I had felt before in the sixteen years since fragile X first entered our family's life. I cursed the gene, I yelled at God, I wept to think that three generations of my family now were under the yoke of the senseless microscopic march of a line of nucleotides.

Words fail me when I try to describe the horror of watching my younger sister clamp a washcloth in her mouth to keep it closed. She couldn't chew, so she lost fifteen pounds, subsisting on yogurt and liquids, including the occasional martini to dull the pain. Through it all, Cathy was stoic, which made it all the harder to watch. If it had been me, I know I would have lost my mind, but she bore it all as if it were her lot.

Cathy didn't have children as the rest of us did. One night at a family dinner, Cathy made the astonishing statement—her speech was excruciating but comprehensible—that what she was going through was nowhere as bad as what my sisters and I suffered with our children with fragile X. I was flabbergasted. My father's and my sister's suffering breaks my heart in a way that my son's challenges do not. J.P. is who he always was, born with a syndrome that retarded his development but couldn't repress his joyful spirit or his impish wit. I had made peace with his compromised future. Cathy's suffering was of an altogether different order for me.

Here we were, parsing suffering around the remains of Easter dinner. Who has it worse, the mother of a son who is mentally retarded or a woman who cannot eat or talk? That was a simple one for me. But it was a mark of how low the family trajectory had sunk that we were forced to entertain such questions.

Eventually, Cathy's neurologist in St. Louis prescribed a dopamine drug that he thought was a long shot—only 1 to 3 percent of patients are helped by the drug. Miraculously, Cathy was in that tiny minority and, to this day, has been able to control the dystonia with several daily doses. It seems a miracle that she can talk normally and eat solid food again; we only pray that she doesn't become resistant to the drug, as sometimes happens.

My son, J.P., like most people with fragile X, likes to anticipate the future. Surprises are unwelcome for him, and schedules are a lifeline to security. When you tell J.P. what is going to happen next—a trip to the grocery store, a party, a visitor to our house—he has a habit of asking, "And then what?" You come up with the next activity, and he asks again, "And then what?" as if he's listening to an enthralling story.

A traditional plot moves in a linear fashion, forward into the future, and it has a beginning, a middle, and an end. I thought our story, J.P.'s and mine, would go like this: his diagnosis would be followed by treatment and improvement in his condition, while my grief would give way to acceptance and peace. Progress. Closure. What I passed down to him would end if he didn't reproduce.

Instead, our family story is a retrograde tale, with each generation seeming to pass the ball back up along the line. And there is no end in sight, let alone closure. True to its name, our "dynamic mutation" keeps evolving and changing, just as the science that follows it runs to catch up. The website of the National Fragile X Foundation (www.fragilex.org) started out serving one group of patients and now has expanded its headings to include three awful acronyms: FXS, FXTAS, FXPOI.

J.P.'s favorite question takes on a different meaning for my family these days. What next, we ask, in the chain of disasters let loose by our mutation? I stumble over a word and I wonder if I'm tongue-tied by a nascent dystonia. My feet feel numb in the morning, and I wonder if neuropathy is creeping up. I can't help wondering when the long arm of the gene I carry will catch up with me. I feel like I'm being hunted down, although I don't remember committing a crime.

That's when I need to remember J.P. taking that first walk outdoors alone. Coatless in the chilly air, aimlessly wandering (who knows what thoughts played in his mind?), he was for once unafraid, not a victim of his genetic legacy—just a young man open to what might come his way.

# FIFTEEN

# Lettuce and Shoes

Christine Kehl O'Hagan

..........................................................................................................

It was on an Irish emigrant ship headed for New York that my great-grandmother, Bridget Moore, met my great-grandfather, Michael Galvin. Both of them had apprentice jobs waiting for them. Bridget was apprenticed as a "ladies' maid" and Michael as a plumber, but whether they ever followed through is anyone's guess. They were 9 and 10 years old. What they left behind was poverty, hunger, English rule, the tyranny of the Irish Catholic Church, and, last but not least, their families, including Bridget's unnamed little brother, the one who fell down and "walked funny."

Bridget Moore, Michael Galvin, Duchenne muscular dystrophy—welcome to America.

Guillaume Amand Duchenne was a French physician, born into a family of Marseille seafarers, and an amateur photographer who was intrigued by the muscles of the human face. He observed that "true," genuine smiles use not only the muscles around the mouth but also the muscles of the eyes (the only muscles, by the way, that are unaffected by the brutal disease that bears his name), still called the "Duchenne smile."

After Dr. Duchenne's wife died in childbirth, he left Marseille and moved to Paris, where he developed an interest in a strange "wasting disease" affecting young boys, and he opened a clinic for them. It was there that the doctor invented a "harpoon" that pierced the boys' skin, extracting muscle tissue for study—thus trading, I would imagine, the "Duchenne smiles" for "Duchenne screams."

Dr. Duchenne's patients died in adolescence, when their hearts or lungs failed. The discovery of the Duchenne "dystrophy" that bears his name was met with mixed reviews.

"I thought humanity to be afflicted with enough evils," a colleague reportedly told him. "I do not congratulate you, sir, for the new 'gift' you have given it."

In our family, it's the gift that keeps on giving.[1]

In the 1950s, scientists discovered that DMD—like hemophilia—is a "sex-linked" disorder, carried by females and inherited by males. A decade later, scientists thought they could identify the carriers by elevated blood levels of a muscle-burning enzyme called creatine phosphate kinase. Later, they realized that CPK levels fluctuated and were therefore unreliable. Finding out who was a carrier and who wasn't (and this pains me so now, with those baby boys—my brother, Richie; my son, Jamie; my nephews, Christopher and Jason—so trusting and the stakes so high) was hit or miss.

My great-aunt, my grandmother, my mother, my sister, and myself—carriers all.

Mostly "hit," not so much "miss."

My mother, Helen M. Doyle, was born in New York City's Hell's Kitchen, one week after the signing of the Armistice. Though the Great War was over, the Spanish Flu was in full swing, brawling its way through the crowded tenements, spilling out into the streets where not even the horses were spared, their dead bodies piled high in the gutters.

The fourth child and only girl, my mother (named for her blowsy, free-spirited Aunt Helen, called "Nellie"—a dubious honor) was a complete surprise to her middle-aged mother and father, to whom parenthood had not, thus far, been kind. They'd lost their firstborn infant son to meningitis; their second and third sons, ages 8 and 6 when my mother was born, were, in the language of the times, "unwell," a word commonly used back then to describe anything from consumption to menstrual cramps. Like my mother's first cousins, two little boys who lived in the tenement upstairs, Mom's brothers were crippled by a strange, progressive "muscular apathy" that doctors knew little about.

During their early years, all four little boys, Mom's two brothers and her two cousins, seemed healthy enough, but then, little by little, ruined muscle by ruined muscle, none of them was strong enough to walk, go to school, leave the house (the stairs, always the stairs), or, eventually, even to sit up straight. Imagine how it was for Mom's brothers, spending their days tied into sturdy wooden chairs strong enough not to tip over, watching my infant mother investigate the miniscule world of the family's cold-water flat, the baby girl crawling under the table, in and out of the chair legs, pulling herself to her feet, while her brothers, barely able to move, bound to those chairs, watched. She learned to walk by going from one brother to the next, or so her mother had told her, grabbing their knees to steady herself, their voices all the encouragement they could offer. (When my toddler mother swallowed a threaded needle she found lying on the floor, the screams of her brothers saved her.)

In February 1922, when my mother was 3, an auburn-haired, gray-eyed little girl with a soft Irish face and a nub of a chin, a sudden wave of pneumonia (a threat to anyone unable to move, especially in the days before antibiotics) crashed over the Hell's Kitchen brownstone and swept the boys, all four of them, away.

Four boys from one family, gone in a single week, a week that I suspect set the tone for my mother's entire emotional life, the panic attacks and unrelenting anxiety, the fainting that, in her Irish way, she laughed at, thinking we children wouldn't notice her clenched hands, her fingernails digging into her palms, the traces of blood she left on the newspaper's "funny pages," a kind of hysterical stigmata.

She said she didn't remember her cousins, and as for her brothers, she didn't remember *them* so much as she remembered searching for them, on one night in particular, the house filled with company and a woman she didn't know holding her by the shoulders, stopping her from going into the apartment's "front room," where the woman said Mom's brothers were "asleep." My mother told me that she remembered getting out of bed, tip-toeing to the front room, and opening the door just wide enough to peek inside, but there were no brothers there, only the same brown oilcloth on the floor, the same old chairs, the same old scratched-up piano in the corner of the room that her parents had somehow inherited from two old Irish bachelors who had lost their saloon.

When my mother was in her seventies, she admitted that she'd been a high-strung, demanding child, prone to terrible tantrums, foot stamping, fits of pique, refusing to wear a dress or a "waist" (blouse) if it had so much as a trace of stain on one sleeve, all the while knowing in her heart (and feeling guilty) that she was upsetting her parents, who had—as the relatives kept reminding her—been through so much. But maybe it was simply a matter of making enough noise to be heard.

She was the only child left in a tenement filled with adults, and yet when she fell from the top of the brownstone's stoop and landed, face first, on the sidewalk, it was a neighbor who picked her up and washed away the blood. She never told her parents or other relatives what had happened to her, and even though her nose was misshapen and obviously broken, nobody asked.

But I think if she were asked, she'd say that although she had always missed the brothers she never knew (no one to walk home from school with, her only-child's life filled with after-school lessons), hers had been a more or less happy childhood. Her apartment had always smelled of vegetable soup and Kirkman's Yellow Soap, and on winter mornings, before she got up, her mother warmed her clothes in front of the coal stove. The doors in the tenement were never locked, and her aunts—full of gossip and laughter—started their days at Mom's kitchen table, filling her head with stories so irresistible that my mother hardly touched her oatmeal and was always late for school. It was through my mother's eyes that I saw my grandfather sitting in the front room, reading, the sunlight coming through the drawn, rose-colored satin drapes seeming to set his red hair on fire.

But what my mother remembered most were the summer afternoons she spent with her family on the "Irish Riviera" of Rockaway Beach, at a place called Flaming Mamie's, where the fiddles played and the hard shoes stomped, and my grandmother, who had eased three little boys into the earth, whirled around the dance floor so fast that her hair came undone, the hairpins skittered across the floor, and her thick black locks fell to her waist like a velvet curtain.

*Monday comes after Sunday*, my mother wrote in a letter addressed to my sister and myself, shortly before she died, when Pam's family and mine were struggling so hard to keep going, as my son, Jamie, and my nephews, Chris and Jason, were all afflicted with Duchenne's and all settled for the rest of their lives into wheelchairs. *In the midst of your grief, your personal horror*, she wrote, *people laugh and cry and eat and read the newspaper, they buy lettuce and shoes, they wash dishes, they do the laundry, they get caught up in the days and nights and the simple joy of living.*

*They go on, girls, and so will you.*

The walls of my grandmother's apartment were beige, either by choice or by neglect. Wall sconces that never worked sat in stucco squares outlined with curved plaster molding. The living room faced the rear of the building and was always dark. There were two windows softened by sheer, yellowed curtains, and on each windowsill stood an ugly snake plant in a green flowerpot that my grandmother watered by standing it in the filled bathroom sink, believing as she did in watering from the roots upward. (This must have worked—the plants survived my grandmother.)

The sofa was a worn green brocade, the chair where my grandfather once sat was also brocade, maroon and saggy-bottomed. In one corner of the living room was a tweed chair that was so rough that when I sat in it, it always bit the backs of my thighs.

There was a floor lamp with a yellowed silk shade, two white lamps shaped like loving cups on the end tables, a mahogany dining table that seated twelve, folded up and set against the far wall. Most everything was a cast-off. My mother, young and energetic, was always on the lookout for ways to spruce up the apartment, especially after my grandfather died and my grandmother fell into a sadness that she never left.

On Saturday mornings, my mother sent my father off in our aqua and white Ford to some stranger's apartment or house, to pick up this or that. Once it was a heavy metal-framed bed, then a rocking chair, then an old-fashioned secretary desk, then a double-door white enamel stove. It was through things that my mother tried to keep the old ladies interested in life, but it never worked. The living room always felt barebones, and it smelled—in the winter at least—of weariness and dust.

But in the warm months, the three of us (my sister, Pam, my brother, Richie, and I) loved nothing more than sitting on the windowsill in my grandmother's bedroom. Right outside the window was a garden thick with purple lilacs and yellow forsythia whose blossoms looked like the fingers of a lady's glove. They were blooms we couldn't get close enough to, pressing our noses against the dusty green screens so hard that we came away with tic-tac-toe grids, like Lenten ashes, on our foreheads.

When our parents went out somewhere, usually to a church dance, and the three of us spent the night, we slept in our grandmother's "high bed," as Nana called it, Pam and I on either end, Richie in the middle. Nana and Aunt Nellie put chairs around the bed so none of us would fall out, chairs that looked like prison bars, the mattress so old and soft that sleeping there felt like smothering or drowning, and it was only the loud ticking of the living room clock that kept me tethered to the here and now. The passing cars tossed rectangles of light onto the ceiling that ran down the wall and disappeared. The sexy crack of high heels on the sidewalk outside Nana's bedroom window and the low laughter passing by hinted of secret things between men and women that I somehow already knew and, at the same time, couldn't imagine. Richie slept between the two of us, rolling against me, or rolling against Pam. He was soft and small, and that was when I first worried that something was wrong with him. He walked funny and he fell down. A lot. Nobody seemed to know why.

It is no exaggeration to say that Duchenne muscular dystrophy filled the rooms of that old apartment, just as it was about to fill the rooms, days, hours of our lives.

When I was 5, I found an unframed, sepia photograph stashed on the back of a shelf in my grandmother's coat closet. Two little boys, the older one had his hair cut straight across the forehead and my mother's sad eyes, the younger one with full cheeks and a head full of curls.

"Those were my brothers," my mother said, and then she went back to peeling potatoes or vacuuming the living room rug, the incessant details of her careful housekeeping.

When I was 7, my mother was pregnant again. It was going to be a girl, my mother said, she just knew it, and we would call her Michelle. When the doctor called to tell us that the new baby was a boy, my grandmother screamed, burst into tears, ran out of the room. My father looked at me, I looked at him. He'd wanted to celebrate. Shaken, he handed me and Pam two tiny glasses of beer.

Richie didn't walk until he was 14 months old, and when he did, he moved slowly and deliberately. He could neither run nor climb stairs, not even curbs. When he fell down—and he fell down a lot—he turned white with fear. It was his big, dark, uncomprehending eyes looking up at me from the sidewalk, or the living room rug, or wherever he lay sprawled, that sent me into a terrible despair, an 11-year-old child beset with anxiety and fear, failing in school, failing at life, day after day after day with the taste of soot, metal, ash upon my tongue, while my sister, at 7, hid silently behind the dolls piled on top of her bed. Richie fell and fell some more, bruises upon bruises, until his legs looked as though they were clustered with Concord grapes.

And all we were told was that Richie had "some trouble with the stairs."

When I saw how quickly Richie was getting worse, I skipped the denial phase of grieving and went straight to bargaining. I'd been so indoctrinated by Catholic school and by the nuns' stories of the self-denying, self-mortifying saints guiding Roman soldiers' swords to their throats, burning themselves at the stake, the "all for Jesus" suffering that seemed to be so effective, that in my child's mind, I thought fixing Richie would be a simple matter of taking him to Mass on Saturday mornings, too, and asking God to fix him.

And just to show God that I was serious, I wore the most self-denying clothes I could dig up: a green spangled hat, an ugly plaid jacket someone had given my mother, a bulky white cable-knit sweater with thick purple stripes running down the front and fringes at the bottom. Wearing this outfit was as close to self-flagellating as I could get, being not in ancient Rome but in Jackson Heights, Queens.

Whatever it took to fix my brother, I was willing to do. I was certain that God would grant me this favor, reward me for my efforts, and in a month or so of taking him to Saturday morning Masses, Richie would miraculously become another kind of kid, blooming with activity, healthy and robust just like all the other little kids on our street—but to my great disappointment that never happened. When my brother and I sat in the pew, I held his hand and kept my eyes squeezed shut, praying so hard that my breath felt like an interruption, but it didn't work. God, forever mute, simply turned His face and looked the other way. Richie couldn't stop falling. By the time Richie was 9—the same age as with my son, the same age as with my nephews—he was in a wheelchair. As I read somewhere later, God doesn't work on people time, it's the other way around, and the answer to most prayers is no.

I try not to think about Richie much, gone now these thirty-three years, although sometimes, it's the weirdest thing, how, in my dreams, I can feel his watermelon-smelling hair brush my lips. I'm not aware that I'm thinking about him, and then, it's a cold day and I'm in front of Kmart watching my 11-old-granddaughter stop in her tracks to button her 4-year-old brother's jacket, tie his hood tight, pull his collar up against the wind.

*Brother*, I think, and the word is foreign on my tongue, part of another language.

Had my younger son, Jamie, been as strong and healthy as Patrick, my older boy, I might have led a different life. I might have stayed at the bank where I was a teller-in-training and become head teller, or maybe, at the IRS, I would have been promoted from mailroom clerk to mailroom supervisor, or the manager of the collections department in the investigative firm where I started as a typist. But of course, I'll never know. Despite Jamie's blooming, glorious early years (just like my cousins, my uncles, my brother, Richie, and eventually, my nephews Chris and Jason), Jamie had DMD. I walked away from every job I ever had whenever Jamie was sick, and he was sick a lot, the dark cloud of a pulmonary crisis looming at the end of every hallway. The terrible, awful, unbelievable risk we took in having Jamie was the same risk we'd taken only two years before when we had Patrick, who was beautifully, gloriously healthy. Youth, stupidity, innocence, recklessness, the air-headed certainty that Jamie, like Patrick, would be fine—all of that—but what it all came down to, really, was that we were so in love with our first baby son that we couldn't wait to have another. How could anything go wrong the second time when the first time was such a success?

Jamie was diagnosed in an inner city office by a pediatric neurologist whose sorry beige office was filled with special toys and plenty of space for wheelchairs. There were bars on the windows and vertical blinds that formed a grid of pain I was desperate to escape. DMD took so much from all of us, but it gave me writing and saved my life.

I became a writer because my son was dying and I didn't want to die with him. It was to save myself that I put myself into a novel that I thought had nothing to do with Jamie or DMD.

My first novel was the story of a woman I knew who had delivered a stillborn baby girl and, in her grief, turned to alcohol. "Man takes a drink," my mother said, "and then drink takes the man." I assume that goes for women as well, for when the drink took this woman, it was relentless. She became a barfly, picking up one stranger after another, until one night she took home the wrong man and he strangled her. It was a terrible story I couldn't forget, the trajectory from strangled daughter to strangled mother took ten gruesome years.

The heroine of my novel, however, faced with the same circumstances, survives, and I felt omnipotent, as if I'd given the murdered woman another chance.

But when I gave my first reading in that crowded, hot bookstore, and looked down at Jamie, sitting in front of the podium in his wheelchair, looking paler and sicker than ever before, I realized in a truer, deeper way that he was going to die, and that the heroine of my novel had nothing to do with the murdered woman. I thought that she needed me to tell her story, but it turned out that I needed her more. She was simply a character that I'd created to save my future self, someone I'd sent on ahead, and she was waving to me from a distant shore.

After our sons died (Jamie in 1998, Jason in 2006), Pam and I turned their bedrooms, not into "shrines," as some bereaved parents do, but into "studios" inside which we tried to incubate, like donor surrogates, the artistic talents (writing, painting) that DMD had denied our sons. There we sat, my sister and I, in two different studios, in two different states, eight years apart, waiting for art—as everyone promises—to come save us. Art is capricious, though, stealthy, works undercover, has its own timetable. When it comes to art, the best you can hope for is to be there when it shows up.

"Show your sister how" was what our mother had told me countless times when Pam jumped into the whole teenage Kotex, make-up, dating scene, and back then I was wealthy with big sister advice. But after Jason died, I couldn't tell her how to grieve. I was heartsick when she followed me through that long, black tunnel, praying to God that she'd emerge at the other end, disheveled and sweaty perhaps, yet grateful to be alive. Grief is the original do-it-yourself project, and the secret to getting through is following all the "yeses," no matter how small, from the toothless grin of some baby in a stroller who catches your eye to those autumn days of heartbreaking beauty. You have to keep the yeses in sight.

Not everyone makes it through, and when your child dies, it's a constant struggle not to fall back in, a continual battle with the abyss. There you are, going along fine, off like the rest of the world, singing along to the car radio, buying lettuce and shoes, and then when you're stopped at a traffic light, next to a school bus, you look up—and the child's hand pressed against the dirty window stops your breath. Or there's a certain snippet of music you hear that hits you in places you didn't know were exposed, in the way that ice cream finds a cavity in an unsuspecting tooth.

When my sister describes DMD's latest assault on my nephew, Chris (the "outlier" at age 36), bright shiny yeses be damned, it stops my breath. I'm in awe of DMD's lasting power and continual rage.

Jamie had deep-set blue eyes, caterpillar-thick eyelashes, freckles, and although he'd been auburn-haired as a baby, as he grew older, his hair turned dark. Although he was 24 when he died, Jamie had a child's perpetually runny nose, a dandelion soft beard, a wispy moustache—there were things about Jamie that never quite took root—but his face, I knew, his handsome face, was etched forever on my heart.

Then one day, Patrick Jr., who lives nearby, came to our house to borrow the big ladder. He got out of his car, walked up the driveway, and stood in the sunlight on the porch. He looked different to me, though I didn't know how.

There was something unusual about his face.

Finally, it hit me.

"What happened to your chin?" I asked him.

"My chin? What happened to it?"

"Your scar," I said, "it's gone."

"That was Jamie," Patrick almost whispered, "Jamie had the scar in his chin."

I remembered. Patrick, age 9, scraping his 7-year-old brother off the driveway and carrying him, on his back, into the house, the two of them covered in blood.

*Forget*, I tell myself, *forget*.

There have been so many crying conversations across so many luncheonette tables between my sister and myself about the horrors of DMD that

I wonder what other middle-aged sisters talk about. I feel so bad for her. I have a healthy son, a wonderful daughter-in-law, two gorgeous grandchildren, and my sister does not. Sometimes I feel like the last person to step on the elevator. I turn around, and there's my sister, waiting on the platform for whatever, for her, comes next. There are empty spaces around my lost brother, my lost son, my lost nephew, into which I, as a writer, often imagine different futures, other pasts. My Jamie as tall as his father and his brother, another good husband, another good dad; my nephews, blond giants like their father, striding through the world, making their mark. Chris is the only one left now, his drive to live humbling to everyone else. He cannot move, he cannot breathe on his own, his voice has been stilled to less than a whisper, and yet he lives his life like anyone else. It's the blessed dailyness that takes him—like you, like me—from one small hour to the next. He watches TV shows and movies and surfs the Web—the lettuce and shoes of life—yet his failings are still terrifying, and his recoveries, for all of us, are miracles. It's as if Richie, Jamie, Jason are all contained in Chris's small body, like Russian nesting dolls.

"You can't come with me," is what Jason, in the hospital, told my sister on the last day of his life.

"I'll find you in the stars," she promised, and who knows, maybe that's where all the lost boys wait.

When Jamie was still alive, after Pat Jr. had gone onto his own life, I imagined that someday it would just be the two of us, Patrick and myself, sitting at the beach side by side, sitting together on the LIRR to Manhattan, kneeling in the yard planting flowers, gardening gloves, his 'n' hers.

What I didn't realize was how much older we would be when the time came or how much losing Jamie would take out of us.

We're gray-haired now, grumpy and lumpy (and sometimes, sleepy and sneezy too) and in a youth-oriented culture, completely invisible, something our contemporaries complain about, but to us—after all those years of wheelchairs, ramps, clamps, and spectators, half-expecting our arrival anywhere to be accompanied by buglers, helpless to do anything but follow the lead of our tough Irish son, wise-cracking and laughing our way through it—invisibility is vastly underappreciated.

It's been fourteen years since we carried Jamie anywhere, our arms trembling with the effort. Fourteen years since we prayed him, Lamaze-like, into one more breath, then another, and another after that, hoping that this time, the answer to our prayers would be yes. Fourteen years since we spoon-fed him in restaurants where not everyone looked away, or even here at home—praying again and again that he wouldn't choke.

For fourteen years, we've been getting up from the sofa at night, turning off the TV, putting ourselves to bed, and miracle of miracles, sleeping through until morning. There are no more Darth Vader sounds in the pitch dark coming from the breathing equipment, no more alarms, no more fractured-sleep fights as to who got up last. There's no more panic when the bi-pap breaks down, who cares if there's a blackout?

When I was younger, I thought that our family had been persecuted by some hundred odd years of DMD, singled out unfairly, suffered so much more than any other, but as I grew older, I realized that although our family's struggle was very hard, others have had it worse. I might have learned earlier than most that the clichés are true: everyone loses, you're not alone, laugh as much as you can, hang onto everyone you love. (I would say "dance in the rain," but there's a word count here.)

As the Irish proverb puts it, "It is in the shelter of each other that people live."

But the price that we paid for our present quiet life, our peaceful existence, and even the lessons we learned along the way, no matter how precious, no matter how true, was much too high.

Jamie, we hardly knew you.

# SIXTEEN

# Dear Dr. Frankenstein

*Creation Up Close*

Emily Rapp

...........................................................................................

How can I describe my emotions at this catastrophe, or how delineate the wretch whom with such
infinite pains and care I had endeavored to
form? . . . The different accidents of life are not so
changeable as the feelings of human nature. I had
worked hard for nearly two years, for the sole
purpose of infusing life into an inanimate object.
For this I had deprived myself of rest and health. I
had desired with an ardour that far exceeded
moderation; but now that I had finished, the
beauty of my dream vanished, and breathless
horror and disgust filled my heart. Unable to
endure the aspect of the being I had created, I
rushed out of the room.

VICTOR FRANKENSTEIN

As a woman with a lifelong disability, when I was pregnant I had
every genetic prenatal test available to me. Although my disability is not genetic and my life has always been rich and full and worth
every ounce of its living, it has also involved a great deal of suffering,
both physical and mental, and it was the prospect of this that I wanted to
spare any future child of mine. Even as I felt the miracle of creation—that
stunning pause, a suspension of disbelief—the first time I saw my child's
heartbeat on the ultrasound, I knew that if a compromising genetic
anomaly were discovered, I would have an abortion. I had a battery of

blood tests and a chorionic villus sampling procedure, and each time, waiting for results, I braced myself for loss.

During the CVS procedure, I vividly remember watching the skinny tube extract the smallest bit of fluid from the sac around my child, its sex still unknown; the little microscopic bundle squirmed and twisted although I couldn't yet feel the movements in my body. A few months later I received 3-D ultrasound results, and it was then that I fell fully in love. A face had formed, the thin line of a hand, eyes, feet, a cat-like nose. Not just a microscopic image, but a baby. A boy. And he was mine. By then, when he moved in my belly I could feel him; my ribs stretched, my stomach swelled. I read countless books about how to be a good parent, and I spun elaborate fantasies in my head of what my child would like to do, what he would look like, which college he would attend. I had so many desires for his future, for the life I'd had a part in creating. At the time I was rereading *Frankenstein*, a book I always revisit when I'm beginning a new project (I was working on a novel). It seemed appropriate to be reading Mary Shelley's novel, with its emphasis on the accidents associated with any act of creation, while Ronan kicked and formed in my belly.

Nine months after his birth, my son, Ronan, was diagnosed with Tay-Sachs disease, a rare genetic neurological condition that is always fatal and mistakenly thought to affect only children of Ashkenazi Jewish descent (Ronan's father is Jewish but I am not). This disorder has no known treatment or cure. Children with this disease develop to the six-month level and then gradually regress into a vegetative state, and without medical interventions they will die before the age of 3. They are, quite literally, slowly uncreated. I had taken the Tay-Sachs blood test offered in the standard prenatal battery of tests, but as I later learned, this test detects only nine of the hundred possible mutations of the disease, and it turns out I had a rare variation of an already rare mutation that was detected only after the sequencing of my DNA. Since the initial test came back clear (or as clear as any blood test can be), my son's father and I waived the Tay-Sachs test at the CVS procedure where it would have been detected. At 35, already an "advanced maternal age," my worries were focused on Down syndrome and other disorders. I clearly remember the genetic counselor telling me, "If you're not Jewish, the odds are astronomically small that you are a Tay-Sachs carrier." But the odds, it turned out, didn't matter. The son that I loved would be taken from me in the most severe and heart-wrenching way. I would lose him. I would be forced to watch his decline, his drop into unconscious existence, his seizures, his gradual paralysis and blindness, his suffering, and finally his death. Now 2½, he can no longer move or see. He will never speak. At some point very soon he will stop being able to eat, and then he will die.

On the day of his diagnosis, my duties as a mother immediately shifted from the desire to launch my child into the future to (I would argue) the equally primal desire to make his life as full of dignity as possible. I was determined not to make the mistake the good doctor in Mary Shelley's novel made; I would not abandon my creation. I had so desired to give birth to my child, to be a mother, yet now I found myself gripped by another desire that brought with it a difficult, wrenching grief: I wished I'd never had him.

This moment was a kind of annihilation, as I struggled to make intellectual sense of the many feelings I was having. These conflicting desires—wanting to save Ronan while wishing he'd never been born—touch upon an essential element of creativity, part of which involves experiencing the boundaries of the self dissolving. This proximity to self-dissolution drives both grief and love and makes both interdependent components of the creative process. We think, mistakenly, that it's about control. Dr. Frankenstein's desire to control what he had made is what genetic testing is all about; the desire to control the outcome, control the experience, the attempt to control possible loss. But although he resists the lesson, Dr. Frankenstein finally learns in the most painful way possible that loss is always snuggled right up to joy.

In November 2010 I was at Yaddo, an artist's colony in upstate New York, feverishly trying to finish a draft of my novel before Ronan was born. I was frantic, believing everybody's warnings that "you'll never have time to do anything once you have kids." (This turned out, of course, to be absolutely false, and just another silly thing people say, like "enjoy it now!" or, alternatively, "it's only downhill from here," when you are pregnant and buying groceries at the store, filling the car with gas, trying to live your life.) I went a bit feral at Yaddo, typing away in my little writing room, which was a sun porch clearly better suited for summer residents, with its three walls made entirely of floor-to-ceiling windows. Eventually I admitted that the austerity of the cold was not assisting my creative process and decided to ask for a space heater from the kind caretaker who resembled Walt Whitman. I wore long underwear beneath corduroys and a wool sweater, my heavy down coat, and pink fingerless gloves knitted by a friend (perfect to type in!). One morning, just before dawn, when I hadn't spoken to a human person in nearly three days, hadn't eaten a single meal that did not involve peanut butter, and hadn't slept more than two hours a day in nearly a week—resorting to restless, almost violent naps full of vivid and labyrinthine dreams—I felt my stomach muscles begin to shake and then *move apart*. My ribs started to ache; when I touched them they were electric, ropey wires of vibrating bone, and they, too, were on the move. Muscles cracking, bones stretching. Ronan. The two of us were alone in that room; you probably could have seen the lights in our windows from a long way off.

That morning—the one I think of as the morning of Ronan's growth spurt—there were a few skinny deer nosing around in the scattering of snow outside my window, unimpressed with my artistic ambitions. The trees were cocooned in ice. A terrified-looking squirrel slid ungracefully down a glassy branch and then scurried out of sight. I could see lights on in the cottages across the path; other writers were awake and writing. Ronan kept kicking and elbowing, and I felt my stomach swell beneath my palms. I thought that Ronan must have heard his name mentioned hours before and had decided to make himself known, to take up a bit more space. *Here I am.* I said out loud, "I see you." And later, I did. Checking my email before dinner, I saw that Ronan's father had sent me the digital photograph of my 3-D ultrasound results. I took a breath before I opened the file. There he was, our little Zoat, as we had already nicknamed him, with a face that looked like it was still being molded and shaped, a baby in process, clay-like and soft and sepia-toned. He looked, frankly, like a slightly misshapen cat in mid-meow. But there was a clear nose, eyes that looked squeezed shut, the thin angle of a hand near his mouth. I looked down at my stomach. How had the clay changed from the time that photograph had been taken to now? Had those little webby hands I saw on the screen been pushing at my ribcage? What new parts had been stitched together? Another resident was sitting next to me when I gasped and said, "Look! That's my son! He looks like a malformed cat!"

Isn't he cute?" She craned her neck to get a view of my screen. "Wow," she said. She put a hand on my shoulder. "Thanks for showing me this. I'll never forget it." I could hardly believe he was mine.

At Yaddo, when I reread Frankenstein, it was mostly Mary Shelley I thought about. Some speculation surrounds the origin of her story of a "modern Prometheus." She was obviously familiar with the Greek myth, and the book's epigraph, from *Paradise Lost*, references the creation of Adam, but I imagined that the initial creative spark was a bit more complicated. Mary Shelley's mother, Mary Wollstonecraft, author of the hugely influential early feminist text *A Vindication of the Rights of Woman*, had died giving birth to her daughter. The text she left behind was itself a kind of ghost story, a nightmare tale. And Mary herself would not have been immune to her own private horrors surrounding creation.

One element that draws me to *Frankenstein* is the accidental way the book came to be written. Mary Shelley was in Geneva when it happened—a city I know well, and where I also lived in my early twenties. While summering there, Mary and her friends experienced a weird burst of freakishly cold and nasty weather. As legend has it, in order to pass the time (no *Real Housewives of Zurich* or *French Riviera Shore* for these astute literary minds; they had to make do with imagination!), the group sat inside around the fire and read one another German ghost stories, and then they each agreed to write their own. Mary, 18, was the only one who finished hers. It was published anonymously the following year, in 1818.

Years ago, when I arrived in Geneva to work at an international aid organization just after college, I imagined Mary Shelley arriving in the same place in the same season: the hot, bright summer with its occasional bursts of rain.

My own apartment in Geneva, on the Avenue du Soret, was an enormous, light-filled space snuggled into a concrete building worthy of any Communist apartment bloc. The smell of croissants and espresso floated through my windows. I liked to smoke cigarettes out the kitchen window and watch the cat in the apartment across the alley stalk through the plants lined up on the windowsill as if he were a tiger in a jungle, tracking prey. Once, my neighbor was shaking out her tablecloth from the window when she accidentally let it go—*merde!*—and it floated down the side of the building and landed on the head of a man who was just mounting his motorcycle. We laughed and laughed. "I will bring it to you!" the man shouted up, as if he were saving the woman from a burning building, and a few moments later I heard the bubble of a teapot and the sound of glasses, the two neighbors sharing coffee. I could smell the sharp scent of it through the wall.

In this spacious place (the apartment even had a hallway and a gigantic bathtub, my two favorite parts), my rent was covered (I never knew exactly what it was), my salary was high (I never knew the exact amount, just that I never kept track of what I was spending, as usual, only this time I never ran out of money), and I traveled all over the world for work and made friends with people who themselves were from all over the world. It was a magical, disorienting, identity-forming time. I frequently fell in love and then just as quickly fell out of it, visited Mont Blanc, hiked in the Jura mountains, drank a lot of cheap French wine from big jugs and ate a lot of stinky cheese, chatted up other expatriates at trashy expatriate bars that seemed to exist for that purpose alone, and made one of the best friends of my life, another Emily, who was quickly christened "Other Em." I was 22.

So I could imagine Mary et al. in a spacious apartment in the old city, windows covered with heavy drapes (drafts were potentially dangerous in the 1800s!), walls hung with oil paintings of hunting scenes, dusty tapestries (more hunting scenes, but with different dogs), and maybe a turret or two: nineteenth-century swank. Bells would peal from local churches, but austerely, in a very Calvinist way, befitting the city's religious history, and the smell of chocolate and coffee would still manage to permeate the air, even when the rain fell hard and for days.

Geneva, in essence, seemed a place where all accidents—of appearance, of lifestyle, of difference—were willfully removed. At times it felt as if people had injected some kind of conformist serum: the same outfits, the same ways of speaking, the same food for lunch. I found the conformist spirit weirdly comforting, and also disorienting, The *Genevoise* seemed intent on avoiding chaos, both inside and out.

People, air, public transport—all were clean and efficient. As Other Em once said, "I can't wait to get back to London and see a bloody *dirty* bus." My downstairs neighbor once knocked on my door and demonstrated how I should walk across my floor so that she wouldn't be able to hear my footsteps above her head.

So this is the place where I pictured Mary (Mary Godwin at the time) writing her epic tale, chilling out by the fire with her lover, Percy Shelley, Lord Byron and his pregnant lover, Claire, in odd summer weather described politely in some accounts as "ungenial." The seasons have gotten confused, and these brainy literary types are fascinated with current scientific attempts to "animate" dead matter—a movement called, appropriately, "galvanism"—and from this convergence of factors the urge to write ghost stories springs.

In later accounts, Mary claimed to have written *Frankenstein* in a "waking dream." I had plenty of those at Yaddo. Dreams in which falling rain became thick ropes of hair, then rubber, then iron bars. I could navigate the morphing rain but could not escape it. I imagined Mary writing her supernatural tale in such a state, not yet knowing that these intellectual and philosophical discussions about the animation of dead matter would mean something much different to her, years later, as she survived (or did she?) the death of three children. When writing became her only comfort, and perhaps a poor one at that.

After Ronan's diagnosis, I remember thinking, If there is a single phrase I never want to hear again in my life, it would be "I'm sorry." No doubt Mary Shelley heard "I'm sorry" a lot: from her married lover, from friends and family who helped her bury her dead children. Imagine the response to this letter from Hogg, a friend she wrote to after the death of her first child, a premature daughter:

> My dearest Hogg my baby is dead—will you come to see me as soon
> as you can. I wish to see you—It was perfectly well when I went to
> bed—I awoke in the night to give it suck it appeared to be sleeping so
> quietly that I would not awake it. It was dead then, but we did not
> find that out till morning—from its appearance it evidently died
> of convulsions—Will you come—you are so calm a creature &
> Shelley is afraid of a fever from the milk—for I am no longer a
> mother now.[1]

A woman, a writer, a creator (and, I would argue, still a mother), speaking here, from inside a tunnel of grief, a tunnel that in those early months after Ronan's diagnosis, in 2011, I was still learning to navigate, learning its exact size and shape and levels of darkness and how these shift from moment to moment. I longed to go back in time and appear as a ghost who miraculously lights a candle in the middle of the night at Mary's bedside, or show up in one of her waking dreams bearing a magic wand, like that luminous fairy queen in *The Lord of the Rings* who gives Frodo a special little stick to use "when all other lights have gone out." Or maybe I'd just play John Lennon's "Imagine" on a constant loop from an invisible sound system, the song that was often playing that winter on the stereo as I sat writing by my own fire.

In the months following Ronan's terminal diagnosis, I would sit him on my stomach while he laughed and lunged at my face. Noses were a particular favorite, although the holes in ears were also appreciated, and lips were an endless source of amusement. Fingers? Amazing. He would soberly study me for long moments, as if he knew something I did not— even before his terminal diagnosis he was a philosophical dude—before breaking into a wide, wet smile that was more like a silent laugh. He would have been a great silent film star. During the week after Ronan's father and I received the diagnosis ("Babies with Tay-Sachs can live three years with good care," the doctor intoned), I had often been afraid to be alone with Ronan, terrified of the sadness, helplessness, anger, and fear that touching his head or his hands or his face might provoke in me, these jagged feelings that pierced my day and made it hard to do activities as simple as boiling water for tea or pulling a few squares of toilet paper off the roll. These lines from Gerard Manley Hopkins's poem about grief constantly came to mind: "O the mind, mind has mountains; cliffs of fall / frightful, sheer, no-man-fathomed."[2] But on some mornings I held Ronan without crying. And on other mornings I waltzed him around the room without catapulting into the future. So there was progress. At least in the morning.

During those months after Ronan's diagnosis I read *Frankenstein* again and again. Unbelievably, we did not have a copy on any of our bookshelves (probably one of many books stashed in the Wyoming storage unit, that windy and vast graveyard for books), so I checked it out of the local library and found it just as creepy and disturbing and wonderfully melodramatic as I had almost two decades before.

Now, though, my responses to the novel were different. It struck me that ultrasounds and genetic tests—moments from modern parenthood, modern creation, designed in part to eliminate the chance associated with any act of creation—provide a kind of foresight not so dissimilar to the moment when the fictional Dr. Frankenstein meets the creation he has earnestly endeavored to form. When this creation reveals himself to be something other than what the creator wanted or expected—in essence, a monster—Frankenstein flees the scene in a fugue of horror and shame. In this move, Shelley highlights the problematic nature of creation: once we've made something, we're attached to it. In this attachment dwells the roots of our salvation and our despair, and both of these depend on that other primal human impulse that drives the act of creation in the first place: desire. As the poet and philosopher Anne Carson reminds us: "Desire is not simple"; people choose to procreate and have children based on all kinds of desires, often conflicting. Prometheus can always be rebound, often at the creator's own peril. I know this well. He is an ambitious man, obsessed with science and creating "natural wonder." He beavers away at his creation, literally stitching him together (and here Mary *must* have been thinking of the psalmist who thanks God for being "fearfully and wonderfully made").[3] But instead, the monster looks funny. This experimentation with "the unhallowed arts" did not work out the way he'd planned. The monster is not little-baby cute. He has yellow skin and bulging veins and is twice the size of a normal man. Frankenstein flees from him, frightened and horrified.

The rest of Shelley's novel is a dramatic caper complete with Dickensian moments of coincidence, with the creator actively seeking to destroy his creation, and vice versa. An innocent child is murdered and the wrong person is blamed. Frankenstein falls sick, is nursed back to health, and is eventually imprisoned, picking up a long-suffering wife along the way. He stumbles into much of this trouble in part because he refuses to listen to his creation, who is pursuing him—initially, at least—with a single, relatively simple goal: to be loved and acknowledged. The monster, who is never given a name but is identified in the text as "wench" or "it" or other equally pleasant monikers, learns about human love by sitting in the woods, literally set apart, watching a human family giving and receiving tenderness, experiencing grief and anger, moving through various stages of life and relationship. When he finally reaches out to them in the spirit of human connection, he is violently and mercilessly rejected. Hurt, alone, and super pissed off, he heads out with a new mission: to find his creator and then kick his ass for abandoning him in a world where he finds only rejection.

I thought about all of this as the doctors talked about what was coming for Ronan: seizures, total loss of movement, lung suctioning, nebulizers, a nursery full of loud medical equipment. He would not get better; he would not be comfortable. He would live for a while with difficulty, and then he would die. I began to feel impatient with Dr. Frankenstein for abandoning his creation so easily, for being so weak-minded that he couldn't find beauty in a situation that defied his egotistical and self-serving expectations.

Back when I first read *Frankenstein* in junior high school English class, I identified with many of the characters and the emotions in the book—what young reader doesn't love that heady experience?—but I was also deeply uncomfortable that I identified more closely with the monster than with Frankenstein. I was the new kid in a new school in a new state. Each month we drove six hours to another state where I had adjustments made to a wooden leg that leaked and creaked and gave me disgusting sores that were sometimes so painful that I had to grit my teeth when I walked between classes. I was poked and prodded by a creepy prosthetist whose skin reminded me of a wax statue, and then in the hospital I was X-rayed and examined and sized up. The process felt monstrous and made me feel freaky. (Going through TSA checks each time I fly is not such a dissimilar experience.)

We happened to be reading *Frankenstein* during the basketball section of gym class. I could not run, but I had to take gym in order to graduate, so the teacher devised this brilliant solution: At the end of every session, I would be called out onto the court, where I would stand at the free throw line, a little bit unstable—my artificial feet were literally foam blocks then—as the rest of the class threw balls at me, one by one, and collectively counted how many I could sink. I spent thirty minutes sitting on the bleachers, shivering with anticipation as I prepared for this task, trying to imagine myself out on the court, nailing twenty perfect free throws: *swish, swish, swish.* As it turns out, I did sink a lot of those balls. My dad, a high school all-star player at his small high school in rural Illinois, shot hoops with me in the driveway during the summer. My mom won the free-throw competition at her fiftieth high school reunion. I always was a good shot.

Still, despite this, I empathized with Dr. Frankenstein. All that work, and then poof! A monster in the house! Not a cuddly baby, not a child prodigy, not even a proper man. A menace, a fiend. Who would want to be blamed for creating a monster that lives to wreak such havoc?

Now, thinking about Shelley's book in the wake of my son's diagnosis, I wanted to tell Dr. F to man up and stop being such an asshole. Be a father already, because that's what you are. You created this being who was actually waiting and worthy of your love, and actually a pretty nice guy until you treated him badly, and then you abandoned him because you were scared and unprepared for the randomness and chance that is a part of every creative act, and because his future was dangerous and un-predictable, and because he wasn't a small, delicate-fingered, to-the-manor-born scientist with a fancy Swiss pedigree, and because people were no doubt going to look at him and judge him and then judge you. Get over it. Because havoc happens on its own, with or without your clever machinations. Stand by your man, you cowardly ninny, even if he has greasy yellow skin and a big head and has to crouch down when he walks through doorways. Understand that when you die, it will be the man you made and tried to destroy who will weep over your grave, his wish granted but his heart broken. *Oh, Frankenstein! Generous and self-devoted being! What does it avail that I now ask thee to pardon me? I, who irretrievably destroyed thee by destroying all thou lovedst? Alas! He is cold; he may not answer me.*[4]

After Ronan was diagnosed, many people shared with me that because all parents want their children to be perfect, they had trouble imagining my feelings about Ronan's condition and, I guess, about Ronan. They wondered why Ronan's father and I weren't both tested for the Tay-Sachs gene, not knowing the facts, as if we'd given no thought at all to starting a family. The comments made me angry and seemed to be more about the people who made them rather than about Ronan, tinged with pity, which is useful to nobody, rather than compassion, which is always in short supply and needed by everyone. These comments were no doubt made in a spirit of support, and I understood that people have trouble knowing what to say or how to respond to people's difficult or "abnormal" situations. I've been rattling off my life story to strangers in elevators for years when they've asked, "Oh, my. What's wrong with you?" (Quick story about me, my artificial leg, how it happened, etc.) "Oh, uh, I'm so sorry." The phrase I hate most.

It was difficult—maybe even impossible—for me to imagine that Ronan was not, in his own way, perfect, if only because he lived the only way he knew how, or could. He lived his life the only way he could, and there was a great deal of perfection—and rare innocence—in that. He never looked like the other kids; he was alone at the free throw line in almost every respect. There were so many things from which I had no way of protecting him, thoughts that put me right at the thinning edge of sanity. But I did not leave him, like Frankenstein's monster, sitting in the middle of a dark forest, lonely, perched on a log and wishing somebody loved him. Not my boy.

Ronan was mine. I didn't want him to be another, different baby. I couldn't imagine not having had a part in creating him, or not having known him. I did not run away from him or rush out of any rooms. I stayed put.

And I never wanted him to be perfect. I wanted him to *live*.

# Epilogue

Amy Boesky

......................................................................................

I write entirely to find out what I'm thinking, what
I'm looking at, what I see and what it means . . .
what I want and what I fear.

> JOAN DIDION, "Why I Write"

A fter I published my memoir about BRCA1, a friend asked whether
I minded that perfect strangers now know intimate details about
my life. "What about your students?" she asked, looking at me with a
funny expression. "Don't you feel . . . you know, kind of *naked* in front
of them now?"

I thought about this. I reminded her that students don't always (or
even usually) read everything their professors write. But in truth, from
time to time I actually *do* feel kind of naked—or at least, aware that the
decision to reveal something personal is not one that can be reversed.

For many of us who have contributed to this volume, writing about
genetic mutations has become part of the process of self-discovery that
Didion describes in the epigraph above. The "nakedness" my friend asked
me about is one consequence of that process: as we uncover how we feel
about genetic testing, seeking voluntary medical intervention, or exploring reproductive technologies, we expose those feelings—to ourselves, at
least. Writing about them involves both the process of discovery and the
act of announcing it.

Putting our voices together changes, however subtly, how we each
perceive our own stories. Inside individual lives or families, perceptions
grow familiar over time. We know, often too well, the principal characters

involved, our central issues and dilemmas. We know the decisions made and the ones deferred. Even inside the wider "family" constituted by those who share a condition or disease—the BRCA community, for instance—there is a shared vocabulary, certain common habits of mind.

The de-familiarization gained from moving outside these smaller cultures is important. Stepping into the wider world to join other individuals contemplating the meaning and importance of genetics in their lives allows us to see our own situations with fresh eyes. The terms we've grown habituated to are now less easily recognized. Coming together as a group involves a kind of metaphysical and psychological travel, which, like all travel, demands a degree of discomfort even as it enables new insights. Writing for (and at times against) each other's stories, we find our ideas growing more complicated. We are able to see connections across conditions that remind us how urgently such conversations need to be expanded.

As we completed the volume, an early reader suggested that I ask contributors to offer closing comments about what they see as the value of reading these essays collectively. What last issues do they want readers to take away?

First, contributors emphasize the importance of offsetting rapidly expanding genetic "information" with personal stories. Such stories, as Joanna Rudnick notes, matter as much as scientific data:

> As important as our study of the human genome in the laboratory is our ability to talk about what it is truly like to live with [the] risk and disease that mutated genes bring and their impact on self-identity, relationships, privacy, and our psychological well-being. A collection . . . will help individuals like myself—living at the heart of predictive genetic testing—make sense of our new reality and give us further permission to explore it.

Several contributors underscore the challenges of this kind of writing. One admits she was surprised by how difficult it was to return to these issues: "It's tempting sometimes just to forget about these things. I'd forgotten how hard this is." Several repeat how problematic it may be to reveal personal stories while remaining sensitive to other family members. Yet they confirm that they see sharing these stories as an essential part of expanding public knowledge and decreasing lingering feelings of shame, isolation, and secrecy.

"Genetic knowledge strikes at the heart of identity," Clare Dunsford remarks. Yet given the "invisibility" of so many genetic mutations, how can the importance of this knowledge be shared? Remarking on the personal magnitude of genetic conditions, Isabel Stenzel Byrnes expresses concern that the wider culture might not yet understand how (and why) genetic identity matters: "People may not appreciate how being born with a genetic mutation has the potential to greatly influence one's sense of self. I am a person with CF, but this illness—from one tiny mutation in my genes—has had such a profound impact on my identity in terms of how I see myself, how I tackle the world, how I problem-solve, communicate, love and express myself." Reading through a collection of genetic stories, Isabel notes, may help others understand why a "tiny" mutation can have so profound an effect.

Several contributors express their desire that a volume like this may offset the isolation that is so often a part of living with a genetic predisposition or condition. Mara Faulkner, reflecting on the loneliness that can accompany any lifelong disease, hopes these essays may help others see that they are not alone and offer them guidance. Part of combating isolation is helping to break down stigma. Clare Dunsford suggests that the "invisible" quality of many genetic mutations may augment "an irrational shame at the way our essence has been defined as pathology." Michael Downing agrees: "My genuine hope is that a chorus of public voices will help to dispel the shame that often attends a genetic diagnosis (and illness of all varieties)." Patrick Tracey feels it is especially important to bring hereditary schizophrenia into conversation with other known familial diseases, ameliorating some of the intense shame that accrues around mental illness.

As the writers consider the lingering stigma attached to genetic disease, several look back to an earlier era of secrecy and shame with pronounced sadness. Christine O'Hagan remembers asking her mother about a photograph of the "lost boys," stashed in the closet—her mother's brothers who died in childhood from Duchenne muscular dystrophy—only to have her question brushed aside. Alice Wexler recalls the stigma her mother experienced in the decades when eugenics and genetics shared an uneasy alliance. Patrick Tracey considers the challenges of talking openly about hereditary schizophrenia. In my own family, my sisters and I, numb with grief when we learned our mother's breast cancer had metastasized, tried to remember who we were "allowed" to tell she was sick. For many of us, such secrecy can be harder than the condition it seeks to hide.

But how best to reveal these stories? Several writers point out the limited terms available for writing about genetic difference and wonder whether this will change in coming years. Alice Wexler notes the tendency for writers to emphasize numbers in talking and writing about genetic disease: "I've been thinking about how critical numbers have been in relation to HD, the 'number' of CAG repeats, etc. . . . Genetic mutations are often named by numbers as well, like APOE4 and BRCA1 and 2." How, she wonders, might such numeration matter to the genetic "measurement" of self?

Several authors hope these essays—especially when read together—will counter prevailing tendencies to simplify genetic stories, or to assume every medical issue can be repaired. Emily Rapp hopes the collection both "complicates and simplifies the 'subject' of genetics, which has a tendency to get trotted out in conversations as a kind of inevitability or a 'curse' or a disease of some kind; . . . people begin the genetics conversation (and I mean this in general) with the notion that there's a way to 'fix' everything."

The reality that many genetic diseases or conditions can't currently be prevented or cured is the greatest challenge facing many contributors here. There are (clearly) plenty of challenges associated with genetic mutations. But there are a number of positive experiences as well. There is excitement and gratitude for innovations and ongoing research. There is also the profound sense, running through some of these essays, that the line between "health" and "disease" is not as clearly drawn as many would like to believe and that recognizing the unstable nature of both categories may reinforce the recognition that all lives have value. "What kind of life is that?" one contributor was asked about her child, as if to live with a disability meant not to live at all. In fact, many essays here counter stories of challenges with examples of humor, creativity, strength, and tremendous resilience. As Anabel Stenzel notes, while a genetic condition may pose difficulties, "it can also bring many positives—unique life experiences, awareness of the value of life, of what really matters, a deeper spiritual depth, compassion, etc." Anabel's comment is echoed by Jennifer Rosner, who points to the highly personal nature of the "challenges *and* victories for those living with genetic conditions."

Jennifer is keenly aware of connections and sense of intimacy within the family and at the same time feels "the outward thrust toward integrating into the larger world." She thinks often about the "burden of making (and defending) huge decisions for a child, well before the child has any voice of her own." Jennifer also feels that the "value of this volume may reside in how it highlights how very personal the experiences are, to get a real feel for the complexity, intellectual and emotional" facing people living with genetic conditions.

The appreciation of how rapidly genetic information is expanding is noted by several contributors. Anabel Stenzel hopes readers will come away with an appreciation for the universality of genetic conditions— eventually we will learn that "we are all genetically predisposed to something." "Many—if not most—adults have preexisting conditions, whether or not they have been identified," Michael Downing notes. Recognizing this, we need to be better equipped to share this information. Misha Angrist takes this up in reiterating his desire for widespread transparency in disclosing genetic information:

> We stand on the precipice of a world in which access to one's genome will be routine for most people in most developed countries. At the same time, patients and families are increasingly in a position to share information about themselves. Is it not possible to collectively reimagine how we negotiate our relationship to our own biology?

"Dealing with genetic disease—it impacts 'a village' and it 'takes a village,'" Anabel Stenzel observes. "We are not alone in this experience, as it impacts our families, friends, co-workers, communities, health care providers, and all contribute to our survival and well-being. Genetic disease affects us all; we are all in this together."

Kate Preskenis and Christine O'Hagan both remark on the extent to which disease—however individualized—can produce common emotions and challenges. "Each disease is really the same disease," Christine says. "Shock, hurt, pain, the worst in you, the best in you." Kate hopes that this collection can offer a "desegregation of diseases, a coming together to celebrate similarities and to honor differences." The sense of celebration is echoed by Isabel Stenzel, who comments, "With any health conditions, often there is a tendency to compare one's lot with another's. My hope is that each of us can respect and appreciate the diversity of all of these genetic stories."

Finally, several writers point to the responsibility that new information demands. We share a widespread sense of being at the frontier of a new era of personalized genetic information. Clare Dunsford remarks that "those of us living with genetic conditions are at the frontier of science and metaphysics. We are brave adventurers . . . into the unknown, as genetics scurries to keep up with lived experience." At the same time, rapid changes in the field of genetics make it all the more important, as Laurie Strongin observes, to keep our eyes fixed, not just on the clear-cut, black-and-white decisions, but on the in-between "gray area" where so much of human experience lies:

> The field of genetics is changing rapidly. Information that merely years ago was longed for is now readily available, increasing the options available to us with regard to decisions about reproduction and disease prevention and treatment. As new information and procedures become available, new ethical questions are raised. From the outside, it is easy to form opinions and judgments about decisions that other people make, from selecting embryos to save an existing child's life, to having a double mastectomy to prevent cancer that is more likely due to a genetic predisposition. From afar, these and related issues can appear to be black and white and as such, easy to judge. Hearing individual accounts re-places these issues in the gray area where, in fact, they lie. Gaining an in-depth look at real-life experiences makes these complex issues more accessible and easy to deliberate and understand, and can help change minds and policies.

We are aware, confirmed as we are in the importance of these first steps, how great is the distance ahead. In writing, we retrace steps taken, reconsider moments where paths diverged. We do so in large part, I believe, to help us clarify the second part of Didion's comment that opens this epilogue—to find out what we want and what we fear.

What we fear? Stigma. Isolation. Misunderstanding. Oversimplification. Facile judgment. A restricted, limited sense of what "health" is and whose life has value. Shrinking resources; policy decisions that restrict, rather than expand, the choices and opportunities available to all.

And what we want? For a start, we want to expand current conversations about genetics and identity—to deepen debates, generate questions. Not to mandate a single response or point of view, but to get (and keep) people talking about issues that matter. To remind people that these issues are complicated, and dynamic, and deeply personal.

Years from now, as new truths emerge about genetics, new stories will emerge alongside them. Some of us will be around to share those stories, and it will be interesting to look back to where we stood early in the second decade of this new millennium, when excitement and optimism about what comes next is so deeply mixed with uncertainty about what it will all mean—when what we fear and what we want are so complexly and uneasily intertwined.

# Notes

............................................................................................

## Introduction

*Epigraph.* Lennard J. Davis, "The End of Identity Politics," in *The Disability Studies Reader*, 3rd edition (New York: Routledge, 2010 [first published 1997]), 309.

1. In using the term *identity*, I am suggesting not something fixed or essential but, instead, a complex, multifaceted conception of self. For a review of positions on the term, see Anna Mollow, "Identity Politics and Disability Studies: A Critique of Recent Theory," *Michigan Quarterly Review* 43, no. 2 (2004): 269–97.

2. Francis Collins, *The Language of Life: DNA and the Revolution in Personalized Medicine* (New York: Harper Perennial, 2010), 74. Masha Gessen's discussion on risk analysis with Harvard economics professor David Laibson offers a more detailed model for decision making in the face of a BRCA mutation. Masha Gessen, *Blood Matters: From Inherited Illness to Designer Babies, How the World and I Found Ourselves in the Future of the Gene* (New York: Houghton Mifflin Harcourt, 2008), 85–92.

3. Collins, *Language of Life*, 73.

4. For an overview of current risk-reduction options for women with BRCA mutations, see Sue Friedman, Rebecca Sutphen, and Kathy Steligo, *Confronting Hereditary Breast and Ovarian Cancer: Identify Your Risk, Understand Your Options, Change Your Destiny* (Baltimore: Johns Hopkins Univ. Press, 2012).

5. Joanna Rudnick, "The Spirit of Difference," Positive Exposure, www.positiveexposure.org.

6. On the risks of "geneticizing disability," see James C. Wilson, "(Re)Writing the Genetic Body-Text: Disability, Textuality, and the Human Genome Project," *Cultural Critique*, no. 50 (winter 2000): 23–39.

7. Susan Sontag, *Illness as Metaphor* (New York: Farrar, Straus and Giroux, 1978).

8. In *Cracking the Genome*, for instance, Keith Davies talks about the need to identity the "rogue" genes "lurking" in our genomes. Keith Davies, *Cracking the Genome: Inside the Race to Unlock Human DNA* (reprint, Baltimore: Johns Hopkins University Press, 2002), 50. James Watson talks about the need for scientists to locate "culprit chromosomes." See Mary Rosner and T. R. Johnson, "Telling Stories: Metaphors of the Human Genome Project," *Hypatia* 10, no. 4 (1995): 104–29, 117.

9. Rosner and Johnson, "Telling Stories," 111. Judith Roof, in *The Poetics of DNA* (Minneapolis: Univ. of Minnesota Press, 2007), explores changes in the metaphors used to represent DNA, tracing a change from metaphors of grandeur, coherence, and master narrative (especially cartographic) around the time of the mapping of the genome to metaphors of parts lists or inventory (109–12).

10. Sontag, *Illness as Metaphor*, 182.

11. The literature on postmodern memoir and autobiography is extensive. See, for instance, Sidonie Smith and Julie Watson (eds.), *Reading Autobiography: A Guide for Interpreting Life Narratives*, 2nd edition (Minneapolis: Univ. of Minnesota Press, 2010), and Charles Baxter (ed.), *The Business of Memory: The Art of Remembering in an Age of Forgetting* (Minneapolis, MN: Graywolf Press, 1999). For a brief but insightful study of postmodern memoir, see Peter Ryan, "The Postmodern Memoir," *Writer's Chronicle*, Mar./Apr. 2012.

12. Arthur Frank, *The Wounded Storyteller: Body, Illness, and Ethics* (Chicago: Univ. of Chicago Press, 1997).

13. Laura Tanner, *Lost Bodies: Inhabiting the Borders of Life and Death* (Ithaca, NY: Cornell Univ. Press, 2006), 64, 73.

14. See Cathy Caruth (ed.), *Trauma: Explorations in Memory* (Baltimore: Johns Hopkins Univ. Press, 1995).

15. C. Thomas Couser, "Conflicting Paradigms: The Rhetoric of Disability Memoir," in *Embodied Rhetorics: Disability in Language and Culture*, ed. James C. Wilson and Cynthia Lewiecki-Wilson (Carbondale, IL: Southern Illinois Univ. Press, 2001), 85.

16. Couser, "Conflicting Paradigms," 87.


*Two. Driving North*

1. For more on Bateman's and Holtzman's research, see Washington University in St. Louis, School of Medicine, Department of Neurology, http://neuro .wustl.edu/aboutus/facultybiographies/bateman and http://neuro.wustl.edu/research /researchlabs/holtzmanlaboratory/holtzmanpublications.

### Three. Collateral Damage

1. Names and relationships of family members, doctors, and medical facilities have been changed, in some cases, to protect the identities of those involved. Some quotations have been modified for clarity or concision.

2. Communication with Dr. Kenneth Kosik of the University of California, Santa Barbara, in August 2012 confirmed these statistics. For more facts and figures, see Alzheimer's Association, "2012 Alzheimer's Disease Facts and Figures," *Alzheimer's and Dementia: The Journal of the Alzheimer's Association* 8 (Mar. 2012): 131–68, www.alz.org/alzheimers_disease_facts_and_figures.asp. For statistics on EOAD, see Alzheimer's Association, "Alzheimer's News 3/20/2012," www.alz.org/news_and_events_largest-ever_research_grant.asp; Alzheimer's Association, "Genetic Testing," www.alz.org/documents_custom/statements/genetic _testing.pdf.

### Five. The Unnumbered

1. Elias Canetti, *The Numbered* (London: Marin Boyars, 1984), 12. In the play, each person has a number, allegedly engraved inside a locket presented to him or her at birth, but the locket is opened in secret by an official Keeper of the Lockets only after "the moment" of death. Although everyone knows everybody else's expected life span from their name ("Fifty"), only the individual knows how old he or she is at any given time and therefore how much longer life will continue.

2. I am grateful to Gisli Palsson, whose citation of *The Numbered* in his important paper "Biosocial Relations of Production," *Comparative Studies in Society and History* 51, no. 2 (2009): 288–313, called my attention to this play.

3. Alice Wexler, *Mapping Fate: A Memoir of Family, Risk, and Genetic Research* (Berkeley: Univ. of California Press, 1995). As one widely cited 1992 study concluded after a year of follow-up interviews with persons who had been tested, "predictive testing for Huntington's disease may maintain or even improve the psychological well-being of many people at risk . . . Knowing the result of the predictive test, even if it indicates an increased risk, reduces uncertainty and provides an opportunity for appropriate planning." Even with bad news, according to this study, people "may derive psychological benefits not experienced by those who remain uncertain." See S. Wiggins, P. Whyte, M. Huggins, et al., "The Psychological Consequences of Predictive Testing for Huntington's Disease," *New England Journal of Medicine* 327, no. 20 (1992): 1401–5.

4. Wexler, *Mapping Fate*.

5. There is little disagreement that most of those who develop Huntington's have CAG repeats in the range of 40 or higher. A range of 36 to 39 CAG repeats is considered "reduced penetrance" and may result in affected offspring. A range of 27 to 35 CAGs is often characterized as "high normal." CAGs in this range are highly stable and rarely expand in the next generation, though some authors consider this range to be unstable or incompletely penetrant as well. See D. Brocklebank, J. Gayán, J. M. Andresen, et al., "Repeat Instability in the 27–39 CAG Range of the HD Gene in the Venezuelan Kindreds: Counseling Implications," *American Journal of Medical Genetics Part B Neuropsychiatric Genetics* 150B, no. 3 (2009): 425–29; Huntington Study Group COHORT Investigators, "Characterization of a Large Group of Individuals with Huntington Disease and Their Relatives Enrolled in the COHORT Study," *PLoS One* 7, no. 8 (2012).

6. Kimberly A. Quaid, "A Few Words from a 'Wise' Woman," in *Genes and Human Self-Knowledge: Historical and Philosophical Reflections on Modern Genetics*, ed. Robert F. Weir, Susan C. Lawrence, and Evan Fales (Iowa City: Univ. of Iowa Press, 1994), 5–6. According to Quaid, "Information that one has, or is at risk for, a genetic condition is more intensely personal than information about an illness contracted as a result of contact with an external cause, such as a virus. Genetic information is widely viewed as saying something about who the person is at some fundamental, if unarticulated, level . . . In some cases, even medical professionals exhibit the tendency to treat proven genetic disorders in a manner different from the way they might treat other diseases."

7. Maxine Hong Kingston, *The Woman Warrior: Memoirs of a Girlhood among Ghosts* (New York: Vintage Books, 1975), 6.

8. Charles B. Davenport, *Heredity in Relation to Eugenics* (New York: Henry Holt, 1911). On the uses of Davenport's textbook, see Steven Selden, *Inheriting Shame: The Story of Eugenics and Racism in America* (New York: Teachers College Press, 1999).

9. James R. Scobie, "Confidential Letter," Oct. 29, 1971.

10. Alessandro Portelli, *The Death of Luigi Trastulli and Other Stories: Form and Meaning in Oral History* (New York: SUNY Press, 1991), 19.

11. Nancy S. Wexler, "Genetic 'Russian Roulette': The Experience of Being 'At Risk' for Huntington's Disease," in *Genetic Counseling: Psychological Dimensions*, ed. Seymour Kessler (New York: Academic Press, 1979), 218.

12. The genetic marker identified in 1983 was actually a small stretch of DNA located so close to the abnormal gene on chromosome 4 that the marker and the gene were typically inherited together (or "linked"). If you were at risk and you had the genetic marker, you had about a 98 percent probability of also having the abnormal gene, and vice versa. This "linkage test" was the earliest form of predictive testing for HD. However, it required participation from several

family members and was a complicated procedure. Predictive testing became technically much simpler after 1993, when the abnormal gene itself—with its expanded stretch of CAG repeats—was identified.

13. Joan Didion, "Why I Write," *New York Times Book Review*, Dec. 5, 1976.

14. Gordon Solvut, "Genetic Testing Makes an Ounce of Prediction Worth a Pound of Fear," *Minneapolis Star-Tribune*, Oct. 25, 1995.

15. Philippa Levine, "The Culture of Medicine and the Culture of the Academy," *Radical History Review* 74 (1999): 184–96.

16. M. Gargiulo, S. Lejeune, M.-L. Tanguy, et al., "Long-Term Outcome of Presymptomatic Testing in Huntington Disease," *European Journal of Human Genetics* 17, no. 2 (2009): 165–71; Kimberly A. Quaid, "Long Term Outcomes of Presymptomatic Testing in Huntington Disease," PredictER News, http://pre dicter.blogspot.com/2009/02. See also Nancy S. Wexler, "The Oracle of DNA," in *Molecular Genetics of Neuromuscular Disease*, ed. L. P. Rowland (Oxford: Oxford Univ. Press, 1989); W. D. Hall, R. Mathews, and K. I. Morley, "Being More Realistic about the Public Health Impact of Genomic Medicine," *PLOS Medicine* 7 (Oct. 2012): 1–4.

17. Steven Pinker, "My Genomic Self," *New York Times Magazine*, Jan. 11, 2009, 24. Within the past several years, the challenge for those living with knowledge that they carry the abnormal HD gene has become even greater as research demonstrates that they may begin to show damage and loss of volume in the brain more than a decade before any observable symptoms appear. See E. H. Aylward, P. C. Nopoulos, C. A. Ross, et al., "Identification of Neuroimaging Biomarkers in Preclinical HD: Results from Predict-HD," *Clinical Genetics* 80, suppl. 1 (2011): 3. Early signs of Alzheimer's have also been identified in asymptomatic individuals.

18. S. Creighton, E. W. Almqvist, D. MacGregor, et al., "Predictive, Pre-natal and Diagnostic Genetic Testing for Huntington's Disease: The Experience in Canada from 1987 to 2000." *Clinical Genetics* 63 (2003): 462–75.

19. Institut de coproduction de savoir sur la maladie de Huntington, Dingdingdong, www.dingdingdong.org; Emilie Hermant, *Reveiller l'aurore* (Paris: Seuil, 2009).

20. For example, Jeff Carroll, Ken Serbin, Alice Riviere, Matt Ellison, Sally Spaulding, Katie Moser, and Chris Furbee.


*Six. Of Helices, HIPAA, Hairballs . . . and Humans*

1. G. M. Church, "Genomes for All," *Scientific American* 294, no. 1 (2006): 46–54.

2. Misha Angrist, *Here Is a Human Being: At the Dawn of Personal Genomics* (New York: Harper, 2010).

3. F. Casals and J. Bertranpetit, "Genetics: Human Genetic Variation, Shared and Private," *Science* 337, no. 6090 (2012): 39–40.

4. H. P. Green, "Genetic Technology: Law and Policy for the Brave New World," *Indiana Law Journal* 48, no. 4 (1973): 559–80.

5. Richard J. Hernstein and Charles A. Murray, *The Bell Curve: Intelligence and Class Structure in American Life* (New York: Simon and Schuster, 1996).

6. See E. C. Hayden, "Informed Consent: A Broken Contract," *Nature* 486, no. 7403 (2012): 312–14, doi:10:1038/486312a; D. Franklin, "Uninformed Consent," *Scientific American* 304, no. 3 (2011): 24–25.

7. Sharyl J. Nass, Laura A. Levitt, and Lawrence O. Gostin (eds.), *Beyond the HIPAA Privacy Rule: Enhancing Privacy, Improving Health through Research* (Washington, DC: National Academies Press, 2009); E. Friedman, "HIPAA Humdrum: Generally Speaking, Laws Work Better If They're Enforced," *Modern Healthcare* 41, no. 35 (2011): 26.

8. S. F. Terry, "FDA and CLIA Oversight of Advanced Diagnostics and Biomarker Tests," *Genetic Testing and Molecular Biomarkers* 14, no. 3 (2010): 285–87.

9. J. A. Rovner, "Making Sense of HIPAA Privacy: Solutions for Complex Compliance Dilemmas," *Journal of Health Law* 37, no. 3 (2004): 399–427.

10. D. W. Craig, R. M. Goor, Z. Wang, et al., "Assessing and Managing Risk When Sharing Aggregate Genetic Variant Data," *Nature Reviews Genetics* 12, no. 10 (2011): 730–36.

11. E. A. Feldman, "The Genetic Information Nondiscrimination Act (GINA): Public Policy and Medical Practice in the Age of Personalized Medicine," *Journal of General Internal Medicine* 27, no. 6 (2012): 743–46.

12. M. B. Steck and J. A. Eggert, "The Need to Be Aware and Beware of the Genetic Information Nondiscrimination Act," *Clinical Journal of Oncology Nursing* 15, no. 3 (2011): E34–41.

13. Steven Greenhouse and Michael Barbaro, "Wal-Mart Memo Suggests Ways to Cut Employee Benefit Costs," *New York Times*, Oct. 26, 2005.

14. L. Roberts, "Plan for Genome Centers Sparks a Controversy," *Science* 246, no. 4927 (1989): 204–5.

15. Anonymous, "Genetics and the Public Interest," *Nature* 356, no. 6368 (1992): 365–66.

16. R. J. Yaes, "Funding the Human Genome Project," *JAMA* 264, no. 22 (1990): 2866–67.

17. Watson is quoted in Leon Jaroff, "The Gene Hunt," *Time* 133, no. 12 (1989): 62–67.

18. E. S. Lander, "Genomics: Launching a Revolution in Medicine," *Journal of Law, Medicine, and Ethics* 28, 4 suppl. (2000): 3–14.

19. Collins is quoted in K. Weigmann, "The Code, the Text and the Language of God: When Explaining Science and Its Implications to the Lay Public, Metaphors Come in Handy. But Their Indiscriminate Use Could Also Easily Backfire." *EMBO Reports* 5, no. 2 (2004): 116–18.

20. David Stipp, "Biotech Bonanza," *Fortune* 140, no. 9 (1999): 333–34.

21. J. P. Evans, E. M. Meslin, T. M. Marteau, and T. Caulfield, "Genomics: Deflating the Genomic Bubble," *Science* 331, no. 6019 (2011): 861–62, doi:10.1126/science.1198039.

22. M. Perola, "Genetics of Human Stature: Lessons from Genome-wide Association Studies," *Hormone Research in Paediatrics* 76, suppl. 3 (2011): 10–11.

23. A. G. Madian, H. E. Wheeler, R. B. Jones, and M. E. Dolan, "Relating Human Genetic Variation to Variation in Drug Responses," *Trends in Genetics* 28, no. 10 (2012): 487–95.

24. M. Poot, J. J. van der Smagt, E. H. Brilstra, and T. Bourgeron, "Disentangling the Myriad Genomics of Complex Disorders, Specifically Focusing on Autism, Epilepsy, and Schizophrenia," *Cytogenetic and Genome Research* 135, no. 3–4 (2011): 228–40.

25. R. C. Green, J. S. Roberts, L. A. Cupples, et al., "Disclosure of APOE Genotype for Risk of Alzheimer's Disease," *New England Journal of Medicine* 361, no. 3 (2009): 245–54.

26. J. A. Stamatoyannopoulos, "What Does Our Genome Encode?" *Genome Research* 22, no. 9 (2012): 1602–11.

27. E. E. Joh, "Ethics Watch. DNA Theft: Your Genetic Information at Risk," *Nature Reviews Genetics* 12, no. 12 (2011): 808.

28. N. Petrucelli, M. B. Daly, and G. L. Feldman, "Hereditary Breast and Ovarian Cancer Due to Mutations in BRCA1 and BRCA2," *Genetics in Medicine* 12, no. 5 (2010): 245–59.

29. Angrist, *Here Is a Human*.

30. Angrist, *Here Is a Human*.

31. P. Wicks, T. E. Vaughan, M. P. Massagli, and J. Heywood, "Accelerated Clinical Discovery Using Self-Reported Patient Data Collected Online and a Patient-Matching Algorithm," *Nature Biotechnology* 29, no. 5 (2011): 411–14.

32. E. J. Horn and S. F. Terry, "Permission to Share Biospecimens," *Genetic Testing and Molecular Biomarkers* 16, no. 5 (2012): 311–12; M. Swan, "Crowdsourced Health Research Studies: An Important Emerging Complement to Clinical Trials in the Public Health Research Ecosystem," *Journal of Medical Internet Research* 14, no. 2 (2012): e46.

33. A. D. Lander, "The Edges of Understanding," *BMC Biology* 8, no. 40 (2010), doi:10.1186/1741-7007-8-40.

## Seven. The Power of Two

1. For information about the film *Power Of Two*, see www.thepowerof twomovie.com. To register as an organ donor, please see www.donatelifeamerica .net. To support CF, please see www.CFRI.org.

## Nine. "Why Would You Be Wantin' to Know?"

1. The toll of mental illness was so high that, between 1817 and 1961, no other nation had produced, in proportion to its population, so many who were sent to asylums, workhouses, and jails for the conditions we now call bipolar depression and schizophrenia. In the sixty-five years that followed the Great Irish Famine of the 1840s, Ireland's per capita asylum population saw a sevenfold increase, an unparalleled expansion. See E. Fuller Torrey and Judy Miller, *The Invisible Plague: The Rise of Mental Illness from 1750 to the Present* (New Brunswick, NJ: Rutgers Univ. Press, 2002), 152.

2. Schizophrenia has a strong genetic base, but researchers of famine conditions in China and the Netherlands have found that some events in early life, such as maternal malnutrition, could exacerbate a genetic predisposition. A. S. Brown and E. S. Susser, "Prenatal Nutritional Deficiency and Risk of Adult Schizophrenia," *Schizophrenia Bulletin* 34, no. 6 (2008): 1054–63, doi:10.1093 /schbul/sbn096. The Chinese findings are consistent with those of a much smaller Dutch study, which found a nearly two-fold increase in schizophrenia for those born during Holland's so-called Hunger Winter, a war-imposed famine in 1944 and 1945. See D. St Clair, M. Xu, P. Wang, et al., "Rates of Adult Schizophrenia following Prenatal Exposure to the Chinese Famine of 1959–1961," *JAMA* 294, no. 5 (2005): 557–62.

3. See R. E. Straub, Y. Jiang, C. J. MacLean, et al., "Genetic Variation in the 6p22.3 Gene DTNBP1, the Human Ortholog of the Mouse Dysbindin Gene, Is Associated with Schizophrenia," *American Journal of Human Genetics* 71, no. 2 (2002): 337–48. Essentially, there was a chromosomal abnormality for the gene in a statistically significant number of people in Roscommon who were diagnosed with schizophrenia.

## Eleven. Community and Other Ordinary Miracles

1. Ted Kooser, "The Blind Always Come as Such a Surprise," in *Flying at Night: Poems 1965–1985* (Pittsburgh: Univ. of Pittsburgh Press, 1985), 64.

2. Gramsci is quoted in Edward Said, *Orientalism* (New York: Pantheon, 1978), 25.

3. Stephen Rose, "No Boundaries: Retinal Research Thrives in Israel," *In Focus* (publication of the Foundation Fighting Blindness), spring 2012, 3, www.blindness.org/pdf/InFocus/InFocus%20Spring_2012-final.pdf.

4. Vision 2012, the National Conference of the Foundation Fighting Blindness, Minneapolis, June 28–July 1, 2012, www.blindness.org/visions/rewind.php.

5. Todd Finkelmeyer, "Tim Cordes One of Few Sightless Doctors in U.S.," *Cap Times*, June 2, 2012, 5, http://host.madison.conm/ct/news/local/health-med-fit/article.

6. Ryan Knighton, *C'Mon Papa: Dispatches from a Dad in the Dark* (Toronto: Knopf Canada, 2010).

7. Finkelmeyer, "Tim Cordes," 5.

8. Stephen Kuusisto, *Planet of the Blind* (New York: Dial, 1998), 181.

9. Nicholas Kristof, "Journalism and Compassion," *On Being*, radio broadcast, Feb. 9, 2012, www.onbeing.org/program/journalism-and-compassion/transcript.

10. Stephen Kuusisto, "The Beauty Myth," WashingtonPost.com, Nov. 12, 2006, www.washingtonpost.com/we-dyn/content/article.

11. Henry Vaughan, "Night," in *The Sacred Poems and Private Ejaculations* (Boston: Little, Brown, 1854), 244.

12. Kuusisto, *Planet*, 179.


*Twelve. String Theory, or How One Family Listens through Deafness*

1. I follow the convention whereby audiological deafness is spelled with a lowercase *d*, to distinguish it from Deafness with an uppercase *D*, a reference to the culture and community of those with fluency in Sign Language.

2. Center for Jewish Genetics, www.jewishgenetics.org.


*Fourteen. The Long Arm*

1. S. L. Sherman, N. E. Morton, P. A. Jacobs, and G. Turner, "The Marker (X) Syndrome: A Cytogenetic and Genetic Analysis," *Annals of Human Genetics* 48 (1984): 21–37; S. L. Sherman, P. A. Jacobs, N. E. Morton, et al., "Further Segregation Analysis of the Fragile X Syndrome with Special Reference to Transmitting Males," *Human Genetics* 69 (1985): 289–99.

2. Randi Jenssen Hagerman and Paul J. Hagerman (eds.), *Fragile X Syndrome: Diagnosis, Treatment, and Research* (Baltimore: Johns Hopkins Univ. Press, 2002), 114.

3. T. V. Bilousova, L. Dansie, M. Ngo, et al., "Minocycline Promotes Dendritic Spine Maturation and Improves Behavioural Performance in the Fragile X Mouse Model," *Journal of Medical Genetics* 46 (2009): 94–102.

4. A. Healy, R. Rush, and T. Ocain, "Fragile X Syndrome: An Update on Developing Treatment Modalities," *ACS Chemical Neuroscience* 2 (2011): 402–10.

5. P. J. Hagerman and R. J. Hagerman, "Fragile X–Associated Tremor/ Ataxia Syndrome (FXTAS)," *Mental Retardation and Developmental Disabilities Research Reviews* 10 (2004): 25–30.

6. C. E. Schwartz, J. Dean, P. N. Howard-Peebles, et al., "Obstetrical and Gynecological Complications in Fragile X Carriers: A Multicenter Study," *American Journal of Medical Genetics* 51 (1994): 400–402.

7. P. Franke, M. Leboyer, M. Gänsicke, et al., "Genotype-Phenotype Relationship in Female Carriers of the Premutation and Full Mutation of FMR-1," *Psychiatry Research* 80 (1998): 113–27.

### Fifteen. Lettuce and Shoes

1. For information on Duchenne muscular dystrophy, see Alan E. H. Emery, *Muscular Dystrophy: The Facts*, 2nd edition (New York: Oxford University Press, 2000); Paula Johanson, *Muscular Dystrophy* (New York: Rosen Publishing Group, 2008).

### Sixteen. Dear Dr. Frankenstein

*Epigraph.* Mary Shelley, *Frankenstein, or, The Modern Prometheus* (New York: Penguin Classics, 2003), 58.

1. Miranda Seymour, *Mary Shelley* (New York: Grove Press, 2002), 129.

2. Gerard Manley Hopkins, "No Worst, There Is None. Pitched Past Pitch of Grief," in *Gerard Manley Hopkins: Poems and Prose* (New York: Penguin Classics, 1985), 61.

3. Psalms 139:14.

4. Shelley, *Frankenstein*, 221.

### Epilogue

*Epigraph.* Joan Didion, "Why I Write," *New York Times Book Review*, Dec. 5, 1976.

# Contributors

........................................................................

MISHA ANGRIST was among the first (identifiable) human beings to have his entire genome sequenced. He chronicled this experience in his book *Here Is a Human Being: At the Dawn of Personal Genomics* (Harper, 2010). He earned a doctorate in genetics from Case Western Reserve University and an MFA in writing from the Bennington Writing Seminars. His fiction and nonfiction have been published in *Slate, Salon, Nature,* and *Best New American Voices,* among other places. Misha is an assistant professor at the Duke University Institute for Genome Sciences and Policy.

AMY BOESKY teaches literature and writing at Boston College. In addition to scholarship on seventeenth-century British literature, she has published a memoir, *What We Have* (Penguin/Gotham 2010), and personal essays in journals such as *Memoir (and),* the *Michigan Quarterly Review,* and *Gulf Coast.* A 2011–12 recipient of a Howard Grant in Creative Nonfiction from Brown University, she is at work on a study of genetics and narrative.

KELLY CUPO graduated from Boston College in 2011 with dual concentrations in business management and creative writing. For her senior thesis, Kelly wrote a memoir (*Shake*) about her family's experience with genetic illness. She works and travels as an analyst for a management consulting firm, helping to advise clients across the nation on digital technology. She lives in New York and enjoys spending time with her (loud and often singing) family and friends.

MICHAEL DOWNING'S most recent book is a memoir, *Life with Sudden Death: A Tale of Moral Hazard and Medical Misadventures*. His novels include the national bestseller *Perfect Agreement* and *Breakfast with Scot*, which was adapted as a movie that premiered at the Toronto International Film Festival. His nonfiction includes *Shoes Outside the Door: Desire, Devotion, and Excess at San Francisco Zen Center* (a narrative history of the first Buddhist monastery outside Asia), *Spring Forward: The Annual Madness of Daylight Saving Time*, and essays and reviews for the *New York Times, Washington Post, Wall Street Journal,* and other periodicals. He is a frequent commentator on clocks, Congress, and the confusion about daylight saving on NPR, PBS, and network and cable news programs. Michael teaches creative writing at Tufts University. You can read more about his work at michaeldowningbooks.com.

CLARE DUNSFORD is an associate dean in the College of Arts and Sciences at Boston College. Her memoir about her experience as a mother of a boy with fragile X syndrome, *Spelling Love with an X: A Mother, a Son, and the Gene That Binds Them*, was published by Beacon Press (2007). Chapters from the book have appeared in *X Stories: The Personal Side of Fragile X Syndrome* (Flying Trout Press, 2006), in *Love You to Pieces: Creative Writers on Raising a Child with Special Needs* (Beacon Press, 2008), and in the *Kenyon Review*'s winter 2006 issue, which was dedicated to the Human Genome Project. For information on fragile X, please see www.fragilex.org and www.fraxa.org.

MARA FAULKNER, OSB, is a member of St. Benedict's Monastery in St. Joseph, Minnesota. She teaches literature and writing at the College of St. Benedict / St. John's University. She writes poetry and nonfiction prose. Among her works are *Born of Common Hungers: Benedictine Women in Search of Community*, written in collaboration with photographer Annette Brophy, and *Going Blind: A Memoir*, which was a finalist for a Minnesota Book Award in 2010. Her poem "Things I Didn't Know I Loved" won the 2011 Foley Prize given by *America Magazine*.

CHRISTINE KEHL O'HAGAN is the author of the novel *Benediction at the Savoia* and *The Book of Kehls*, a memoir about her family's long struggle with Duchenne muscular dystrophy. A born-and-bred New Yorker, Christine currently lives on Long Island with her husband. She is working on a second memoir, *Lettuce and Shoes*. Her website is www.christinekehlohagan.com.

CHARLIE PIERCE is an acclaimed national essayist and sports writer, as well as a weekly participant in NPR's *Only a Game*. He was a 1996 finalist for a National Magazine prize for his piece on Alzheimer's, expanded into the book *Hard to Forget: An Alzheimer's Story* (Random House, 2000).

KATE PRESKENIS, after losing five family members to genetically inherited Alzheimer's disease, has wondered if she wants to know her own fate. The early death of loved ones combined with the innovative capability of gene testing led Kate to value and share this unique documented history in *The Gene Guillotine: An Early-Onset Alzheimer's Memoir* (www.katepreskenis.com). Kate participated in CNN's *World's Untold Stories* documentary on Alzheimer's, "Filling the Blank" (www .cnn.com/video/?/video/international/2011/01/20/wus.filling.blank .preview.cnn), was invited as a guest with Kerri Miller on Minnesota Public Radio, and was featured in a Q&A for HerCircleEzine.

EMILY RAPP is the author of *Poster Child: A Memoir* and *The Still Point of the Turning World* (Penguin, 2013). Her work has appeared in the *New York Times*, *Salon*, *Slate*, and many other publications. She is a professor of creative writing and English at the Santa Fe University of Art and Design in Santa Fe, New Mexico.

JENNIFER ROSNER is the author of *If a Tree Falls: A Family's Quest to Hear and Be Heard* (Feminist Press, 2010). Her writing has appeared in the *New York Times*, the *Hastings Center Report*, the *Massachusetts Review*, the *Jewish Daily Forward*, *Good Housekeeping*, *Lilith Magazine*, the *Faster Times*, *Wondertime Magazine*, and elsewhere. Jennifer holds a PhD in philosophy from Stanford University and is editor of *The Messy Self* (Paradigm Publishers, 2007). She lives in Massachusetts with her family.

JOANNA RUDNICK is a director and producer specializing in powerful nonfiction storytelling, with a passion for narratives that humanize vital scientific, medical, and environmental topics, provide audiences with alternative avenues to explore pressing social issues, and open up windows into the creative process of our most celebrated artists. For a decade, Joanna has been a filmmaking associate with the award-winning documentary powerhouse Kartemquin Films, the filmmakers who brought us *Hoop Dreams*. *In the Family*, her directorial debut, was nominated for the 2009 Emmy Award for Outstanding Informational Programming—Long Form and was broadcast nationally on PBS's *POV* series in 2008. Most recently, Joanna was the supervising producer on *Crossfire Hurricane*, a documentary film on the genesis of the Rolling Stones, which premiered on HBO in November 2012. Joanna is currently directing *On Beauty*, the story of fashion photographer turned humanitarian Rick Guidotti, who is on a crusade to challenge the way we see the markings of genetic difference.

ANABEL STENZEL and ISABEL STENZEL BYRNES are 41 years old and reside in Redwood City, California, with their husbands. Graduates of Stanford University and the University of California, Berkeley, Anabel works as a genetic counselor at Lucile Packard Children's Hospital at Stanford and Isabel works as a social worker at Sutter Hospice. They lecture around the country and in Japan on their memoir *The Power of Two: A Twin Triumph over Cystic Fibrosis* (University of Missouri Press, 2007) and on personal and professional health topics, and they continue to screen, internationally, their award-winning documentary film *The Power Of Two*. For more information, see www.thepoweroftwomovie.com.

LAURIE STRONGIN is the founder and executive director of the Hope for Henry Foundation, which brings laughter, entertainment, and smiles to seriously ill children. In 2010, Hyperion published her book *Saving Henry*, a memoir about her family's fight to save her son through pioneering developments in the area of genetics and reproductive health. She acts as a family advocate in the national discussion of ethics and genetics. Laurie lives with her husband, Allen Goldberg, and sons, Jack and Joe Goldberg, in Washington, DC. For more information, please visit www.hopeforhenry.org or www.savinghenry .com.

PATRICK TRACEY is a journalist and the author of the PEN award–winning memoir *Stalking Irish Madness*. A contract writer for *Salon* online magazine, he has covered the Occupy movement and the contest for the Massachusetts Senate from his home in Boston. A former contributing writer for the *Washington City Paper*, he has written for *Ms.* magazine, the *Washington Post*, and many other media outlets.

ALICE WEXLER is the author of four books, including *Mapping Fate: A Memoir of Family, Risk, and Genetic Disease* (University of California Press, 1995) and *The Woman Who Walked into the Sea: Huntington's and the Making of a Genetic Disease* (Yale University Press, 2008), which won the American Medical Writers Association Book Award in 2009. She was the recipient of a Guggenheim Foundation Fellowship in 2002 and has received fellowships from the National Endowment for the Humanities, the Social Science Research Council, and the National Library of Medicine. Alice is currently a research scholar at the UCLA Center for the Study of Women and an associate fellow at the UCLA Center for Society and Genetics, and she serves on the board of directors of the Hereditary Disease Foundation. She lives in Santa Monica.

# Index